Physics Practical
for B.Sc. /B.Tech.

THE AUTHORS

Dr. Ramesh Chandra Tiwari, Ph.D is presently working as Professor of Physics, Mizoram University (A Central University), Aizawl, Mizoram, India. He has teaching experience of about 29 years, which includes Post Graduate, Research and Undergraduate levels. He is former Dean, School of Physical Sciences, former Head, Department of Physics (PG/UG) having administrative experience of about 13 years. His area of research interest is Electronics, Thermoluminescence and Physics of ionizing and non-ionizing radiation including Geophysics. He has filed 2 Patents, published 58 papers and 2 books.

Dr. Alok Shukla is an Associate Professor in Physics, National Institute of Technology, Mizoram, Aizawl. He taught Engineering Physics at NIT Mizoram for many years and undertook research in materials science. He has guided 01 student for Ph.D degree and 03 students are working under his guidance. Dr. Shukla has about 30 research papers in national/international journals of repute. Presently, he is working on development of nano-size multiferroic materials under SERB-DST funded research project.

Physics Practical
for B.Sc. /B.Tech.

Ramesh Chandra Tiwari
Alok Shukla

2022
Daya Publishing House®
A Division of
Astral International Pvt. Ltd.
New Delhi – 110 002

ISBN : 9789359191621

Published by : **Daya Publishing House®**
A Division of
Astral International Pvt. Ltd.
– ISO 9001:2015 Certified Company –
4736/23, Ansari Road, Darya Ganj,
New Delhi-110 002
Ph. 011-43549197, 23278134
E-mail: info@astralint.com
Website: www.astralint.com

Preface

The book entitled *"Physics-Practical for B.Sc./B.Tech."* is prepared for the students pursuing graduate degree in Science and Engineering. The objective of the present book is to fulfil the need for a concise laboratory book in Physics, covering the prescribed experiments in a single volume.

The students pursuing science and engineering degree course, spend a fairly good amount of time in laboratory activities. Learning theoretical concepts of physics in laboratory by performing live experiments is a very important aspect of the curriculum in science and engineering. With the help of carefully planned experiments, the students get opportunity to explore the process of experimentation. While performing the experiments, the student develops a habit of inquiry, acquaints itself with hands on experience by using different components, equipments and materials. This leads to learning about the errors, their analysis and interpretation of the results obtained, which finally results in to acquiring skills by the students to handle the experiments more carefully and effectively. Over the time, the student discovers the excitement of learning that is one of the biggest assets of their future career.

Based on our experience acquired by working with students and the problems faced by them in performing the laboratory experiments, the present book is designed to facilitate and to cater their needs. This book describes the experiments covering broad areas of physics namely; General Properties of Matter, Optics, Heat, Electricity and Electronics. We are confident that this book shall be extremely useful to the graduating students of science and engineering streams and we do hope that they will develop a deep interest in laboratory experiments with wider active participation.

We take this opportunity to thank all our colleagues for their valuable suggestions and encouragement. All the efforts have been put to keep the book free from errors, however, any suggestion or comments towards further improvement of this book is highly appreciated and welcomed wholeheartedly.

Ramesh Chandra Tiwari

Alok Shukla

Contents

Preface *v*

1. Band Gap in a Semiconductor Using a Reverse Biased Pn- Diode 1
2. Wavelength of Different Colours by Grating 6
3. Hall Effect Experiment 9
4. Newton's Ring Experiment 12
5. Characteristics of Semiconductor Diode 15
6. Current–Voltage and Illumination Characteristics of a Solar Cell 19
7. Determination of Wavelength of He-Ne Laser 24
8. Determination of Wavelength of Semiconductor Laser 27
9. Series and Parallel Resonant Circuits 30
10. Maxwell's Bridge Experiment 35
11. Refractive Index of the Material of the Prism 39
12. Measurement of Low Resistance 47
13. Transmission Characteristics of a Low Pass Filter 51
14. Principle of Phase Sensitive Detection 55
15. Vertical Torsion Apparatus 60
16. Horizontal Torsion Apparatus 67

17. Young's Modulus Experiment	74
18. Coefficient of Viscosity of Water	80
19. Coefficient of Viscosity of Liquid	85
20. Magnifying Power of a Telescope	91
21. Resolving Power of a Telescope	94
22. Searle's Method	98
23. Jaeger's Method	105
24. Surface Tension of Water	110
25. Poisson's Ratio	116
26. Sonometer	121
27. Planck's Constant	125
28. E/M of Electron	130
29. Ac Measurements	138
30. Npn Transistor	144
31. Pnp Transistor	150

Experiment No. 1

Band Gap in a Semiconductor Using a Reverse Biased Pn- Diode

Aim

To determine the width of the forbidden energy gap Eg in a semiconductor in the form of Germanium pn-junction diode.

Apparatus

The apparatus required for conducting the experiment are: i) dc power supply ii) a heating arrangement to heat the diode (Range 0 - 80ºC) iii) thermometer (range 0 ºC -100 ºC) iv) a micro ammeter and v) a Germanium diode.

Formula

The width of the forbidden energy gap is given by Eg(eV) = 0.198 x slope of the curve.

Circuit

The Figure 1 shows the circuit diagram to perform the experiment. The diode is reverse biased with the help of a 3 volt a dc power supply and the current that flows through the reverse biased diode is measured by using a micro ammeter. A heating system (oven) is connected to increase the temperature of the diode up to 60 ºC

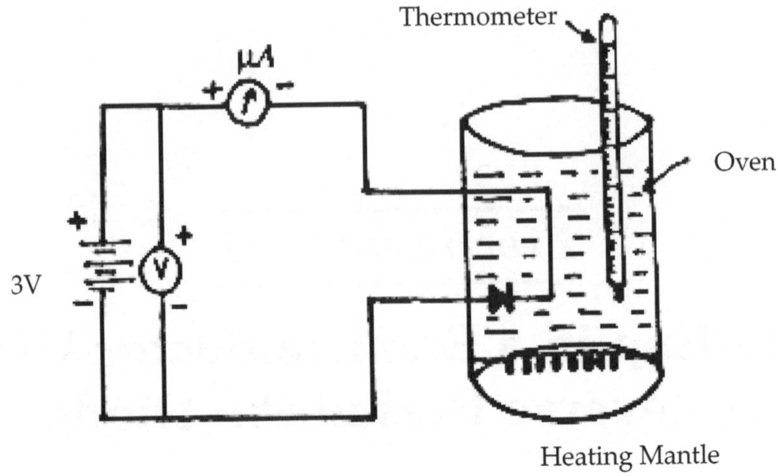

Fig. 1: Experimental arrangement for determination of energy gap.

Principle

The band gap is a very important physical property that characterises the behaviour of a semiconductor material. Energy gap indicates the energy required to free the holes and electrons present for participation in electrical conduction at a specific temperature. Thus the band gap of a semiconductor determines the electrical conduction properties depending on the material used. For further illustration, one can take the example of the band gap of Silicon which is 1.12eV and of Germanium it is 0.72eV. Both values of energy gap are at room temperature. It is clear that at a given temperature the number of free charge carriers available for conduction in case of Germanium will be greater than that of Silicon. Also, the band gap of the material of the semiconductor provides information about the temperature range of the device. Four probe method is one of the most popular method to determine band gap of the semiconductor material available in Wafer form. In this method, one can study the temperature variation of its resistivity. A range of semiconductors are readily available in the form of devices of daily use.

PN junction diode is a very popular semiconductor device, that can be used to determine the energy gap. When a P type semiconductor and N type semiconductor sandwitched together, it results into a PN junction diode. The band gap in a pure semiconductor and the extrinsic semiconductor will be the same. In a P type material the holes are the majority carriers and electrons are minority carriers. Similarly in a N Type material, electrons are majority carriers and holes are minority carriers. When P type and N Type semiconductors form a junction, a space charge region is formed near the junction, known as depletion region or depletion layer. The electric field in the depletion region favors the movement of minority carriers across the junction, while the movement of the majority carriers are prevented by this electric field. It is important to note that the majority carriers in P and N type regions arises due to the impurity dopedin each region, while the effect of temperature leads to generate minority carriers. One can also say that, by supplying energy equal to Eg, covalent bonds are broken, which results into generating minority carriers. Thus one can calculate Eg by studying the temperature variation of the minority carriers. By studying temperature variation of reverse saturation current (under reverse biased condition) one can obtain the variation of minority carriers.

It is given by the relation,

$$I_o ae\left[\frac{d_p}{L_p}Pn+\frac{d_n}{L_n}\right] \tag{1}$$

where, a is the area of the junction, dp and dn are the diffusion coefficients of holes and electrons respectively, Lp and Ln are the diffusion lengths of holes and electrons respectively, pn is the concentration of minority holes in n-type region and n is the concentration of minority electrons in p-type region

Using the law of mass action, namely nnpn = n^2i and pp np = ni^2 and nn = nd and pp = na into equation (1), we get,

$$I_o ae\left[\frac{d_p}{L_p n_d}Pn+\frac{d_n}{L_n n_a}\right] n_i 2 \tag{2}$$

The saturation current is thus propotional to the area of the junction a and ni^2.

But, ni 2 = ncnve-$E^{g/kT}$ (3)

$$I_o ae\left[\frac{d_p}{L_p n_d}+\frac{d_n}{L_n n_a}\right] n_c n_v e^{-E_g kT} \tag{4}$$

or

$$I_o = Ce^{-E_g/kT} \tag{5}$$

Where

$$C = n_c n_v a_e\left[\frac{d_p}{L_p n_d}+\frac{d_n}{L_n n_a}\right]$$

It is evident from equation (5), the saturation current depends on (Eg/kT) exponentially. Taking logarithms on both the sides of equation (5), we get

$$\text{In Io} = \ln C - Eg / kT$$

$$2.303 \log Io = 2.303 \log C - Eg / kT$$

or

$$\log I_o = \log C - \frac{Eg}{2.303k \times kT} \tag{6}$$

or

$$\log I_o = \log C - \frac{Eg}{2.303k \times I_o 3} \cdot \frac{10^3}{T} \tag{7}$$

Equation (6) implies that the reverse saturation current Io in a given semiconductor diode is dependent only on the junction temperature. H varies inversely with temperature. A plot of log Io versus 10^3/T gives a straight line with a negative slope having a value of Eg/2.303k(10^3). In order to determine the slope of the straight line, Eg can be calculated. Thus,

$$[Eg] = 2.303\,(10^3)\,k \times m = 0.198\ m\ \text{(approx.)} \tag{8}$$

Procedure

1. Suitably long and sufficient wires are connected to the diode terminals and the diode is connected into the circuit as shown in Fig.1.
2. The diode is kept in a oven. A thermometer is also kept in the oven such that its mercury bulb is just at die height of the diode.
3. The power supply is switched on and the voltage is adjusted (say 5 volts). The current through the diode and the room temperature are noted.
4. The heating mantle is switched on and the oven is heated up to 60^0 C.
5. The heating mantle is switched off when the temperature of oven reached 60^0C. The temperature of oven stabilises say at 70^0 C.
6. The corresponding current through the diode are noted using microammeter.
7. The oven is allowed to cool slowly. Based on fall in temperature, the current through the diode decreases. As the current falls through steps of 2 μA (or any other convenient interval), the corresponding temperatures are noted down as given in the TABLE – 1.
8. After the completion of calculations are completed a graph is plotted taking $\frac{1000}{T}$ on X – axis and log Io on Y- axis. A straight line is obtained as shown in Fig.2.
9. The slope of the straight line is determined by using Fig. 2, and by using Eq. (8), the bandgap Eg is calculated.

Observations

Room Temperature : ______ ^{0}C, Voltage :______ volts.

TABLE 1

Sl. No.	Current Io (μA)	Temperature t (ºC)	Temperature T= t +273 (^{0}C)	$\frac{1000}{T}$ (K^{-1})	Log Io
1					
2					
3					
4					
5					
6					
7					
8					
9					
10					

Calculations

$$Eg = (0.198 \times \text{slope})\ eV$$
$$= 0.198 \times PR/QR\ eV = \_\_\_\_\_\_\_\_\ eV$$

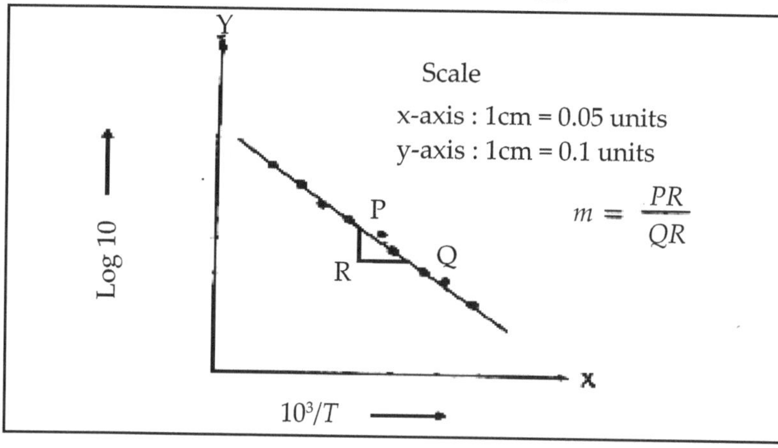

Fig. 2 : Variation of current with inverse temperature in a reverse biased pn-diode.

Results

The width of the forbidden gap (Eg) in germanium semiconductor is found to be ________eV.

Precautions

The following precautions should be taken while performing the experiment:

1. The maximum temperature of the diode should not be raised beyond65°C in order to avoid overheating.
2. Least count of ammeter and voltmeter should be measured accurately.

Viva-Voce

1. What is semiconductor?
2. What is the difference between conductor, insulator and semiconductor?
3. What are the applications of semiconductor?
4. Define band gap energy of a semiconductor.
5. What is n-type and p -type semiconductor?
6. How does the energy gap of a semiconductor varies with temperature and why? Explain the reasons.

Experiment No. 2

Wavelength of Different Colours by Grating

Aim

To determine wavelength of different colours with the help of diffraction grating.

Apparatus

A spectrometer, a grating of known grating element, mercury light.

Principle

When a Parallel beam of light falls normally on a grating most of the light goes straight, part of the light diffracted in different direction on either side of the undeviated light and forms colour and bright bands after passing through a converging lens. First bright band is termed as the first order spectrum and the second bright band is called the second order spectrum. For any spectral lines the wavelength is determined by the following relation.

$$\lambda = \frac{(e+d)\sin\theta}{n}$$

Where (e+d) is grading element,θ is angle of diffraction and n is the order of spectrum.

Procedure

1. Keep the telescope just in front of collimator and see the image of slit on cross wire. This is direct image. Take the reading of both verniers.
2. Rotate the telescope very slowly from this state on either side of it very

3. soon 1st order (n=1) will be seen. Fix up the position of cross wire in the middle of the line of that colour (violet, green etc.) take the reading by both of vernier the difference of and this reading will give the value of angle of diffraction (θ).
4. In the same way the diffraction angle for different colours will be observed.

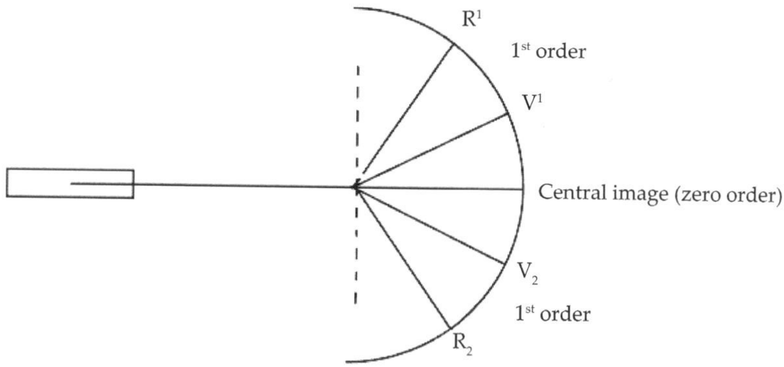

Fig 1 : Spectral lines in Grating

Observation Table

Least count of Vernier =..

Grating element (e+d) $= \dfrac{2.54.}{N}$ ....cms

Sl. No.	Col-ours	Verniers	Order of Spectrum	Reading of Telescope for spectrum						2θ = (x~y)	Mean θ
				Left Image			Right Image				
				M.S.	V.S.	Total (x)	M.S.	V.S.	Total (y)		
	Violet	V1									
		V2									
	Green	V1									
		V2									
	Red	V1									
		V2									

Calculation

Use formula $$\lambda = \frac{(e+d)\sin\theta}{n}$$

Result

The mean wavelength of Mercury light of different colours by grating are:

λ_v = ...Ao

λ_G = ...Aºo

λ_R = ...Ao

Precautions

(i) The grating should always be kept from two sides and in no case the ruled surface should be touched while setting the grating at right angle to the incident light. Two images are seen due to reflection from the two surface of the plate,

(ii) While taking reading of the telescope the grating table must remain clamped.

Viva Voce

1. What is diffraction grating?
2. What do you mean by LPI?
3. What is the difference between prism and grating spectrum?
4. What is grating element?
5. What is the difference between mercury light and sodium light?

Experiment No. 3

Hall Effect

Aim

To calibrate an electromagnet. To study the dependence of Hall voltage on magnetic field and current and to determine the Hall coefficient and carrier concentration.

Apparatus

Electromagnet, Power supply for electromagnet, Gauss meter, Hall probe, constant current source for Hall probe, digital voltmeter(range mV).

Formula

Hall Coefficient $R_H = \frac{V_H t}{iB_x}$, $R_H = \frac{1}{ne}$

Where VH is Hall voltage, t is thickness of sample, i is the current passing through the sample, Bx is applied magnetic field, n is carrier concentration, e is electronic charge or charge on electron (1.6x10^{-19}C)

Diagram

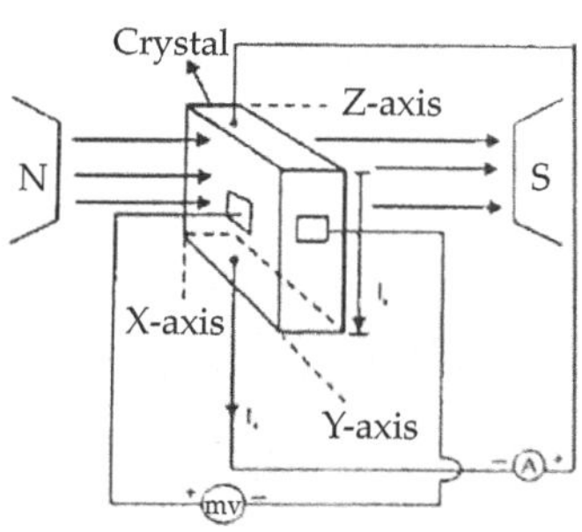

Fig.1 Schematic Diagram for experimental set up .

Procedure

A. For Calibration of Electromagnet

1. Adjust the gap between the pole pieces at 2.0 cm.
2. Adjust the reading of the digital Gauss meter to 0 with the help of Zero adjustment knob when the sensor is away from the electromagnet.
3. Insert the sensor between the pole pieces and switch on the constant current power supply for the electromagnet. Adjust the current through the electromagnet to 0.5 A.
4. Increase the current through the electromagnet in steps of 0.5 A up to 2A. Read the magnetic field in each case.
5. Plot the graph between the current along the X axis and magnetic field along the Y axis. This is the calibration curve for the electromagnet.

B. For Measurement of Hall Voltage.

1. Insert the Hall probe such that the area of cross section of the probe is normal to the magnetic field.
2. Connect the leads to the constant current source as shown in the diagram.
3. Connect the Hall voltage terminals to the digital millivoltmeter(input terminal)
4. Adjust the current through the Hall probe to 1mA. Increase the current through the electromagnet from 0 in steps 0.5 A up to 2A. Measure the Hall voltage each time.
5. Repeat step 4 by changing the current through the sample to 2mA and 3 mA.

Observation Tables

1. For Calibration of Electromagnet.

TABLE - 1

Sl. No.	Current through electromagnet (A)	Magnetic Field (Wb/m^2)

2. For measurement of Hall voltage.

TABLE - 2

Sl. No.	Current through electromagnet (A)	Magnetic Field (Wb/m^2)	Hall voltage in mV Current through sample		
			i = 1 mA	i = 2 mA	i = 3 mA

Calculations

Result

The value of Hall coefficient for a given semiconductor is ....

Precautions

1. The current through the Hall probe should not exceed the specified value.
2. The gap between the pole pieces should be kept constant throughout the experiment.

Viva Voce

1. What is Hall effect?
2. Define Hall voltage.
3. What do you mean by Hall coefficient?
4. What is the use of Gauss meter? Explain 1 Gauss.
5. How does the electromagnet works?
6. What is the unit of magnetic field?

Experiment No. 4

Newton's Ring

Aim

To determine the radius of curvature of plano-convex lens or wavelength λ by Newton's ring experiment.

Apparatus

A convex lens, a glass plate, a wooden box open at the top and at one sided fitted with a glass plate inclined at an angle of 45° with the horizontal, a long focus traveling microscope, sodium lamp, reading lamp etc.

Formula Used

The radius of curvature of lens is given by the formulae

$$R = \frac{D^2_{n+p} - D^2_{n}}{4\lambda p}$$

Where Dn= diameter of nth bright ring with air film, Dn + p =diameter of $(n+p)^{th}$ bright ring with air film and λ = wavelength of sodium light.

Diagram

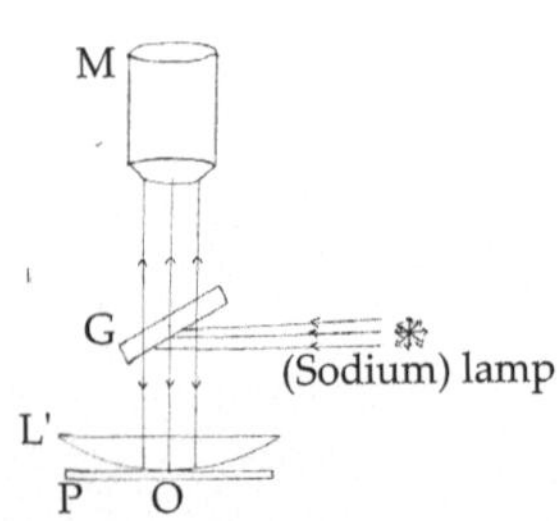

Fig.1 Experimental Arrangement

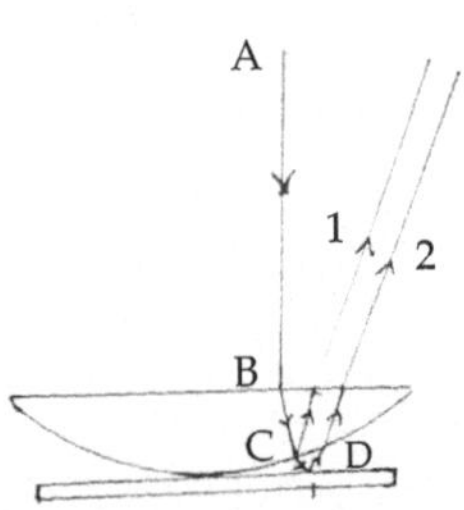

Fig. 2 Formation of Ring

Procedure

1. Place the convex surface of the lens of the plate after cleaning the two surfaces thoroughly and fix them in a frame so that the point of contact is at the centre of the lens and all round this centre wedge shape air film is surrounded.
2. Place the system in the wooden box and illuminate it by parallel rays using sodium light and a convex lens. A screen with an aperture may be used to cut away straight light.
3. Focus the microscope on the surface of the lens. Looking vertically above O and by slightly adjusting the microscope a brilliant set of Newton's rings is observed with central spot dark.
4. Adjust the microscope such that one of the cross-wires is perpendicular to the scale of
5. The microscope and intersection of cross-wires coincides with the centre of dark ring.
6. Slide the microscope towards the left and go on counting till you reach to the 15th bright ring and note the position of cross-wire. Now slide the micro scope to wards right and note down the position of 14th, 13th......6th bright rings. Go on sliding the microscope and start taking readings of positions of bright rings on right of centre right from 6th to 15th. The difference of readings of position of any no. of bright rings on left and right will give the value of diameter of that particular bright ring.

Observation Table

Least Count of the microscope =..............mm.

TABLE 1

Sl. No.	No. of Rings	Microscope Reading		Diameter D=[x-y] (cm)	D^2	$D^2_{n+p} - D^2_n$
		Left end x (cm)	Right end y (cm)			

Calculation

Result

The radius of curvature of a plano-convex lens is............ cm.

Precautions

1. The light from the sodium lamp should be rendered approximately parallel by means of a short focus lens.
2. The lens used for the production of Newton's ring should have surfaces having large radius of curvature. This is necessary for two reasons: [i] the angle θ enclosing the air film becomes very small and [ii] the rings observed have a large diameter and hence the error in the measurement of their diameters is minimized.
3. It may happen that the rings seen are not circular but possess an elliptical shape. This is mainly due to unclean surfaces of the lens and the plate. Therefore in such case clean the lens and the plate with spirit and then set them to get the rings.

Viva Voce

1. What do you mean by Newton's ring?
2. What is the difference between sodium lamp and mercury lamp?
3. What is the standard value of wavelength of sodium light?
4. What do you mean by bright and dark fringes?
5. What is the difference between convex and concave lens?
6. How the light is produced in a sodium lamp? Explain.

Experiment No. 5

Characteristics of Semiconductor Diode

Aim

To determine the forward and reverse bias characteristics of pn junction diodes.

Apparatus

pn-junction diode, Connecting wires, Resistance.

Principle

The typical characteristics of a semiconductor diode are shown in Fig. 1. The mathematical expression showing this behaviour (except in the breakdown region) is given by

$$I = I_0(e^{\frac{qV}{kT}} - 1)$$

In the case of germanium-silicon diodes and light emitting diode the parameters of interest are cut-in voltage Vc, dynamic resistance rd, and the saturation current Io. In a number of devices the reverse saturation current may be too small to be measured. Zener diodes on the other hand are operated in the reverse mode and the parameters of interest include the zener voltage Vz, and the zener resistance rz. All the characteristics mentioned above may be obtained from the experimental unit which consists of (a)A variable voltage supply (b)A voltmeter (c)A set of three precision resistances : 1 ohm, 100 ohm and 10K ohm.

When used with the DVM, these enable current measurements with full scales of 100 mA, 2 mA and 20 μA respectively.

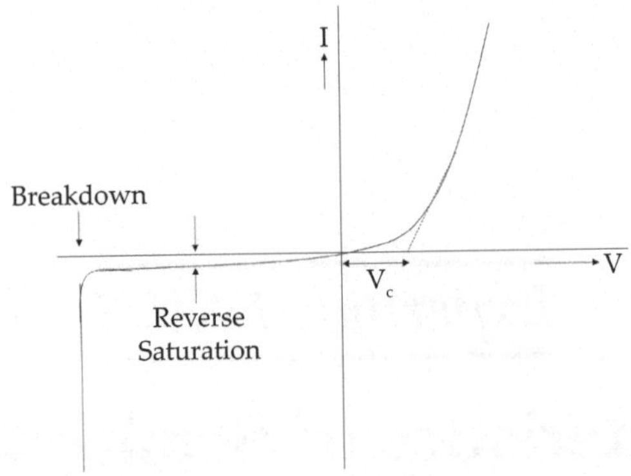

Procedure

A. Forward Characteristics

(a) Set-up the circuit as in Fig. 2 (See layout diagram)

(b) Set the current' limiter suitably, e.g. 100 mA for rectifiers and 20 mA for LED, zener diodes and small signal diodes.

(c) Vary the voltage in small steps and measure the current in terms of the voltage drop across the 1 ohm resistance. Note that while measuring voltage, the DVM has a maximum range of 19.99V (or 20V). During current measurements, DVM is automatically set to 199.9 mV (or 200 mV), and displays the voltage drop across the connecting resistance in mV. The current in mA is obtained by dividing it by the value of resistance. Also, the actual voltage across the diode should be corrected for by taking into account the drop across the current measuring resistance, i.e. VD=VM-VR where VD : Voltage across diode; Vm: Measured voltage; VR: Voltage across current measuring resistance. Tabulate the readings as given below:

Table: 1 (FB)

S.No.	Voltage (V)	Current(mA)

Table:2 (RB)

S.No.	Voltage (V)	Current (uA)

NOTE : To obtain the initial position of the forward characteristics more accurately, the 100 ohm resistance may be used in place of 1 ohm resistance upto a current of approximately 2 mA.

(d) Sketch the V-I characteristics and extend the linear portion of the curve ownward to obtain the cut-in voltage Vc. The slope of the linear portion gives the dynamic resistance rd of the diode.

B. Reverse Bias Characteristics

(a) Set-up the circuit as in Fig. 3.

(b) Set the current limit switch to 20 mA.

(c) Vary the voltage in steps of 1 volt and measure the current in terms of the voltage drop across the 10K resistance. Tabulate the readings.

Note : In a number of devices the reverse current may be too small to be read.

(d) In case of zener diode the reverse current would tend to rise abruptly after the breakdown. To take subsequent readings replace the 10 KΩ resistance QWby 1Ω and continue in small steps of voltage change.

(e) Plot the reverse characteristics and obtain from the graph of the value of reverse saturation current I.

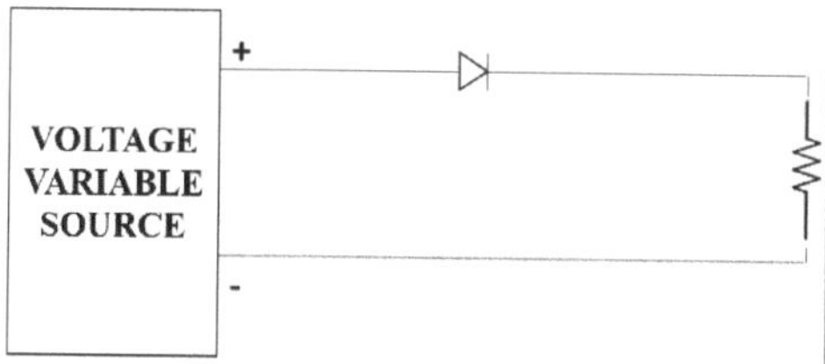

Forward Characteristics

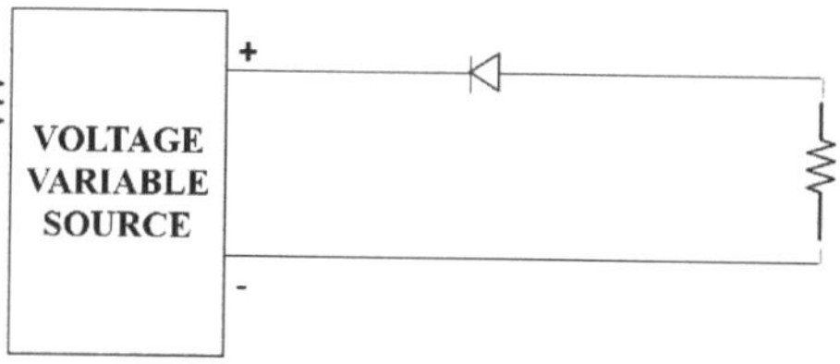

Reverse Characteristics

Result

Forward and Reverse bias characteristics of pn-junction diode has been verified.

Precautions

(a) Set the current limit switch properly. An incorrect setting may damage the device under test.

(b) To sketch the characteristics accurately near the sharp bends (around the cut-in and breakdown points) a larger number of readings may be necessary. Choose suitable resistances, as suggested, for current measurements in these portions.

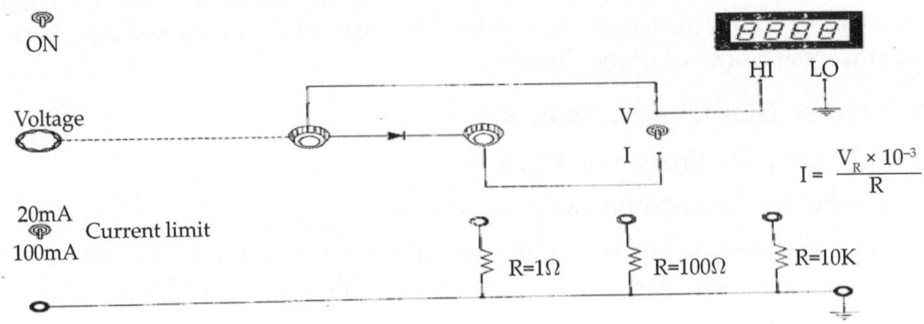

Fig. 4 Circuit diagram

Viva Voce

1. What is a PN junction diode?
2. What is depletion region?
3. Why more voltage is required to conduct the diode in reverse bias condition?
4. What are majority and minority carriers?
5. How semiconductor diode is different than zener diode?
6. What are the applications of semiconductor diode?
7. How diode converts AC in to DC?
8. What is avalanche breakdown?

Experiment No. 6

Current–Voltage and Illumination Characteristics of a Solar Cell

Aim

To study the current – voltage and illumination characteristics of a solar cell.

Apparatus

The apparatus required to perform the experiment are:

i) a solar cell ii) a voltmeter iii) a milliammeter iv) Load resistance.

(v) a 100 W lam pitted with an intensity regulator.

Circuit

Figure 1 shows the circuit for determining the solar cell characteristics. The circuit may be assembled or a ready-to-use training board may be used to conduct the experiment. The solar cell experiment is to be performed using sunlight as the source. It can also be perform using a 100 W bulb.

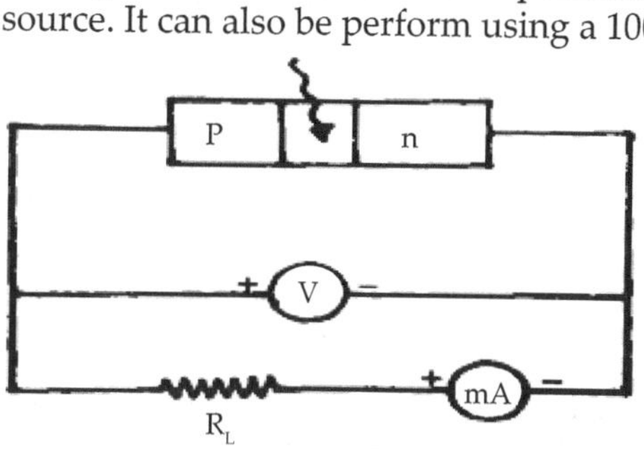

Fig. 1 Circuit diagram

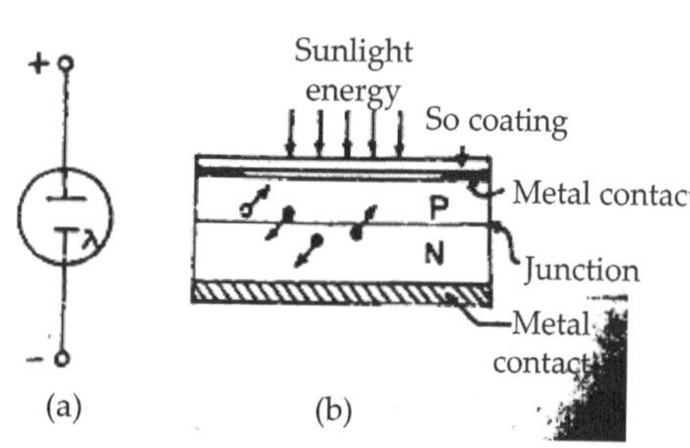

Fig.2 Solar cell circuit symbol& cross-section

Theory

A solar cell is basically a pn-junction device which converts light energy into electrical energy. This phenomenon of conversion of optical energy into electrical energy is known as photovoltaic effect. Due to this reason, a solar cell is also called a photovoltaic cell. When a solar cell is illuminated, the photons incident on the cell generates electron-hole pairs. In the material the electrons and holes reach the junction by diffusion.. At the junction, the barrier potential separates the positive and negative charge carriers. Under the action of the electricfield, the electrons (minority carriers) from p- region move into n-region. Similarly, the holes from n-region move into p-region. When sunlight falls at the surface of the Silicon wafer, which is doped to produce a PN junction, light can penetrate through the PN junction. The energy of sunlight is sufficient enough to break covalent bonds, which leads to the formation of electron- hole pairs on both sides of the junction. The flow of electrons and holes produces a small current whose magnitude depends on the intensity of sunlight proportionally. The current also depends on the surface area of the junction on the solar cell. The electrons and holes accumulated on both sides of the junction produces an open circuit voltage, which also depends on the flux (lumen/cm) of sunlight. An open circuit voltage of about 0.5V can produce about 1 Watt of electrical output power.

The electrons and holes accumulated on both sides of the junction produces an electromotive force called as photo electromotive force (emf). Photo emf is also known as open-circuit voltage. When we connect an external circuit across the terminals of the solar cell, the minority carriers return to their originating sides via external circuit, that causes a current to flow through the circuit. The above described characteristic of solar cell shows that it behaves as a battery with N side as negative terminal and P side as positive terminal. By using a voltmeter, the photo emf or voltage across the terminals can be recorded. In general, solar cells are fabricated by using silicon single crystals, which is opaque to visible light. To solve this problem, the surface layer of the solar cell (P type), is made externally thin so that the maximum amount of light can penetrate the junction.

For a solar cell, it is desired that the absorption of light take place at the junction region such that the electron-hole pairs so generated, can be separated by the junction field, before getting lost by recombination. For enhanced transmission of light into the material P type surface is coated with anti-reflection coating layer vacuum deposited thin metallic films on both sides of the cell, act as electrodes. Generally, the Silicon wafers of 1″ diameter to 4″ are used to fabricate solar cells. A number of PN junctions are formed on a wafer by using a mesh type or finger-like electrode structure, which enhances the total voltage and current output. An array of inter connected solar cells (series or parallel chain) increases the Power Output. The arrangement described above is called a solar panel and a single solar cell is rarely used.

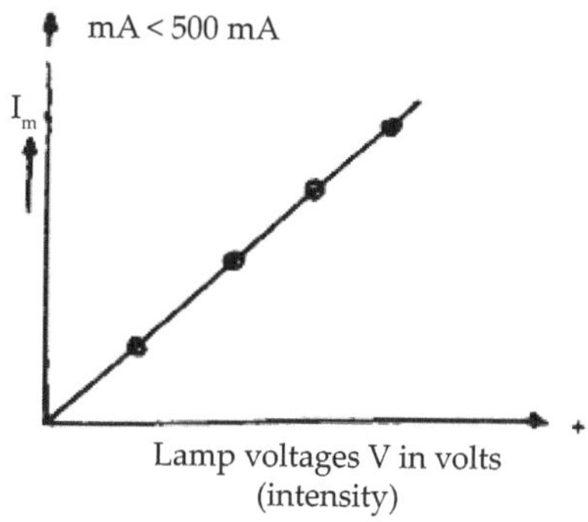

Fig. 3 Illumination characteristics of a solar cell

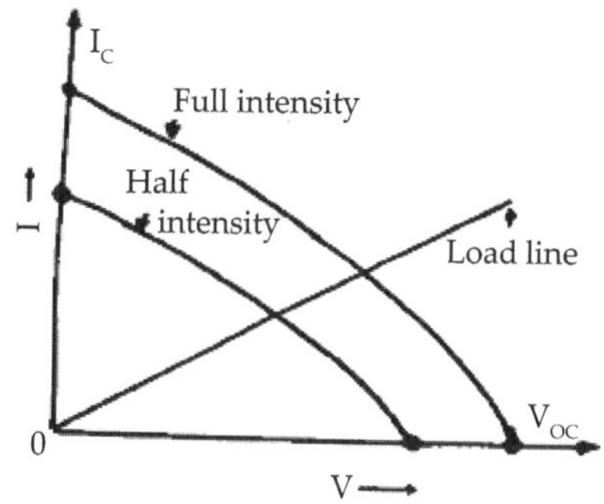

Fig. 4 Load line curve

(i) The illumination characteristics of a solar cell panel are shown in Fig. 3. From Fig. 3 it is evident that the current through the solar cell increases in the same proportion as the intensity of light falling on the panel increases.

(ii) The output current-voltage characteristics of a solar panel are shown in Fig.4. It can be seen that on open circuit, the output voltage of the panel is 0.5 V and the current is zero. If the panel is short-circuited, current is maximum while the output voltage of the panel becomes zero. In both the cases, the output power is zero. It is clear from the curve that the voltage varies depending on the current drawn.

Procedure

A) Illumination Characteristics

1. The circuit is connected as shown in the Fig.1.
2. A 100 W bulb is arranged over the solar panel such that white light falls on it at normal. The intensity control is kept at its minimum and the lamp is switched 'on'.
3. The load is kept at 0 ohms and the short-circuit current is noted down.
4. The intensity of the lamp is varied in steps and the current is noted at each setting. The readings are noted as given in TABLE-1.
5. A graph is drawn between the current and intensity such that the intensity it taken as proportional to the voltage given to the lamp and volts are taken as units on the x-axis in place of intensity.

B) I -V Characteristics

1. The intensity of the lamp is kept at the minimum. The load resistance is not connected into the circuit. The open circuit voltage is noted with zero load resistance is noted down.
2. The output of the solar panel is shorted. The short circuit current is noted.
3. Now, the load key is set at 20 ohms. The voltage and current are noted down. The load is varied in steps upto 10 ohms and at each setting the voltage

and current values are recorded. The readings are entered in TABLE-2 as given below.

4. The intensity of the lamp is increased. At this setting, the load is again varied from 10 to 1000 ohms and the corresponding voltage and current values are recorded. The open circuit voltage and short-circuit current are also obtained and recorded.
5. The intensity of the lamp is set to its maximum and procedure described in step-4 is repeated.
6. A graph for current versus voltage is drawn.

Observations

Illumination Characteristics

RL = O ohms

TABLE 1

Sl. No.	Intensity (volts)	Current (mA)

I - V Characteristics

TABLE 2

Sl No.	RL ohms	Intensity I1	
		Voltage (V)	Current (mA)
1			
2			
3			
4			
5			
6			
7			
8			
9			
10			

Open-circuit voltage (RL= ∞) = __________ Volts for I1

Short-circuit current (RL = 0) = __________ mA for I1

Results

1. The open-circuit voltage developed by the solar panel is found to be .... m volts.

2. The short-circuit current developed by the solar panel is found to be .... mA.
3. Current voltage and illumination characteristics of a solar cell has been verified.

Precautions

1. The terminals of the solar panel are to be correctly connected to the sockets on the training board. Positive terminal should be connected to the positive and negative to negative.
2. The solar cell panel should be used in the short circuit mode for a short period only.

Viva Voce

1. What is photovoltaic effect?
2. What is photo emf? How is it related to intensity of incident light?
3. What do you mean by efficiency of a solar cell?
4. How a solar cell is different from other cells?
5. Explain the principle of working of a solar cell and its applications.

Experiment No. 7

Determination of Wavelength of He-Ne Laser

Aim

To determine the wavelength of He-Ne laser by using transmission grating.

Apparatus Required

He-Ne Laser, Transmission grating, screen and optical bench.

Theory

The wavelength of a laser light can be measured by use of a transmission grating. A transmission grating is an optical component which produces spectrum of light due to diffraction. It has large number of lines grooved on it. If d is spacing between the lines then according to diffraction formula :

$$d \sin \theta = n\lambda$$

Where θ is the angle between the direction and diffractive beam and n is the order of spectrum and λ is the wavelength of laser light.

By knowing the value of d and measuring angle θ subtended between direct beam and first order maxima, we can determine the wavelength of laser light.

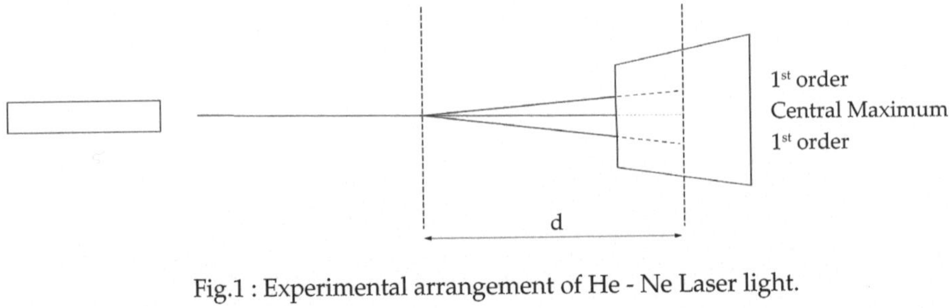

Fig.1 : Experimental arrangement of He - Ne Laser light.

Formula

$$\text{(i) } \theta = \frac{x}{r} (radian)$$

$$= \frac{180 \times x}{n \times r}$$

Where n is the order of spectrum, for first order n=1, and r is the distance between grating and screen.

$$d = \frac{1}{no. of\ lines/\ cm}$$

Procedure

1. He-Ne laser is mounted on its saddle.
2. A plane transmission grating is mounted on the upright next to laser. A screen is mounted next to the grating.
3. The laser is switched on. The relative orientation of laser with respect to grating is adjusted such that the spectral spot is observed on the screen.
4. The screen is moved towards and away from the grating till about four spots are clearly seen on the screen.
5. The central maximum and other maxima are identified.
6. The positions x of the spots belonging to the first order on either side of central maximum are marked.
7. The separation between central maximum and first order maxima is measured on either side. The readings are noted as given in the table.
8. The distance r between the grating and screen is noted.
9. The number of lines per cm is computed.

Observations

TABLE 1

Sl. No.	r (cm)	x (cm)	$\theta = \frac{x}{r}(radian)$	$\theta = \frac{180 \times x}{n \times r}$	sinθ	no. of lines/cm	λ(in Å)
1							
2							
3							
4							
5							

Calculations

$$d = \frac{no}{no.of\ lines/\ cm}$$

$$d \sin \theta = n \lambda$$

$$\lambda = \frac{d \sin\theta}{n}$$

$$\lambda_{avg} = (\lambda_1+\lambda_2+\lambda_3+\lambda_4+\lambda_5)/5$$

$$\lambda_{avg} = \text{....................} \ \text{Å}$$

Result

The average value of wavelength of a given laser light (λ) is =............ Å.

Precautions

(i) Transmission grating should be used carefully.

(ii) Note down the value of d and x with precision.

(iii) Stay away from Laser beam. Do not look at the beam directly.

Viva Voce

1. What is wavelength? How wavelength and frequency are related?
2. How laser light is different than ordinary light?
3. What is stimulated emission of radiation?
4. What is population inversion?
5. In He-Ne laser, what is the ratio of the two gases?
6. Explain the term diffraction and grating.
7. What are the applications of the He-Ne laser?

Experiment No. 8

Determination of Wavelength of Semiconductor Laser

Aim

To determine the wavelength of Semiconductor laser by transmission grating.

Apparatus Required

Semiconductor Laser, transmission grating, screen and optical bench.

Theory

The wavelength of Semiconductor Laser light can be measured by use of transmission grating. A transmission grating is an optical component, which produces spectrum of light due to diffraction. It has large number of lines grooved on it. If d is the spacing between the lines. Then according to diffraction formula,

$$d \sin\theta = n\lambda$$

Where θ is the angle between the direction and diffractive beam and n is the order of spectrum and λ is the wavelength of laser light.

By knowing the value of d and measuring angle θ subtended between direct beam and first order maxima, we can determine the wavelength of laser light.

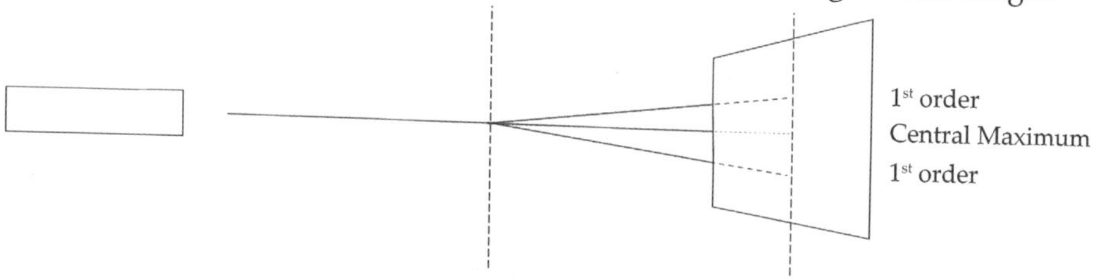

Fig.1 : Experimental arrangement of Semiconductor Laser light.

Formula

$$\text{(i) } \theta = \frac{x}{r}(radian)$$

$$= \frac{180 \times x}{n \times r}$$

Where n is the order of spectrum, for first order n=1, and r is the distance between grating and screen.

$$d = \frac{1}{no.of\ lines/\ cm}$$

Procedure

1. Semiconductor laser is mounted on its saddle.
2. A plane transmission grating is mounted on the upright next to laser. A screen is mounted next to the grating.
3. The laser is switched on. The relative orientation of laser with respect to grating is adjusted such that the spectral spot is observed on the screen.
4. The screen is moved towards and away from the grating till about four spots are clearly seen on the screen.
5. The central maximum and other maxima are identified.
6. The positions x of the spots belonging to the first order on either side of central maximum are marked.
7. The separation between central maximum and first order maxima is measured on either side. The readings are noted as given in the table.
8. The distance r between the grating and screen is noted.
9. The number of lines per cm is computed.

Observations

TABLE 1

Sl. No.	r (cm)	x (cm)	$\theta = \frac{x}{r}(radian)$	$\theta = \frac{180 \times x}{n \times r}$	sinθ	no. of lines/ cm	λ(in Å)
1							
2							
3							
4							
5							

Calculations

$$d = \frac{no}{no.of\ lines/\ cm}$$

$$d \sin \theta = n\ \lambda$$

$$\lambda = \frac{d \sin\theta}{n}$$

$$\lambda_{avg} = (\lambda_1 + \lambda_2 + \lambda_3 + \lambda_4 + \lambda_5)/5$$

$$\lambda_{avg} = \ldots\ldots\ldots\ldots\ \text{Å}$$

Result

The average value of wavelength of a given laser light (λ) is =.............. Å.

Precautions

(i) Transmission grating should be used carefully.

(ii) Note down the value of d and x with precision.

(iii) Stay away from Laser beam. Do not look at the laser beam directly.

Viva Voce

1. How semiconductor laser is different than He-Ne Laser?
2. What is radiative and non-radiative emission?
3. What is coherence of light?
4. Explain the terms spatial and temporal coherence.
5. What are the applications of semiconductor laser?
6. What is threshold gain of semiconductor laser?
7. What are the applications of semiconductor laser?

Experiment No. 9

Series and Parallel Resonant Circuits

Aim

To show that the impedance of a circuit, in which an inductance L and a capacitance C are in series, is a minimum at the resonance frequency fres, and the impedance of a circuit, in which L and C are in parallel, is a maximum at fres.

Apparatus Required

Signal generator, R-L-C box and a Digital Multi Meter (DMM) which can measure frequency up to 10 kHz and AC volts to three decimal places in the range 2 V.

Principle

At the resonance frequency $f_{res} = (1/2\,\pi)\,(1/LC)^{1/2}$, $\omega L = 1/\omega C$. The impedance of an inductance is $Z_L = R_L + j\omega L$ and of a capacitance is $Z_C = -j/\omega C$. When the inductance and capacitance are connected in series the total impedance is $Z_L + R_L$.

This has a minimum value R_L at the resonance frequency. At any other frequency the impedance is more than R_L.

If the coil and capacitance are connected in parallel, the effective impedance of the combination is

$$Z = Z_L Z_C \,/\, (Z_L + Z_C) \qquad (1)$$

At resonance frequency Z_L+Z_C is a minimum. If RL were zero, then $Z_L Z_C$ would be independent of frequency while Z_L+Z_C would be zero at the resonance frequency, making the effective impedance infinite at the resonance frequency. However RL is non-zero and small for the coil. This makes the impedance to go through a finite, but maximum, value at a frequency close to the resonant frequency.

We verify the above results with the following experiment.

Procedure

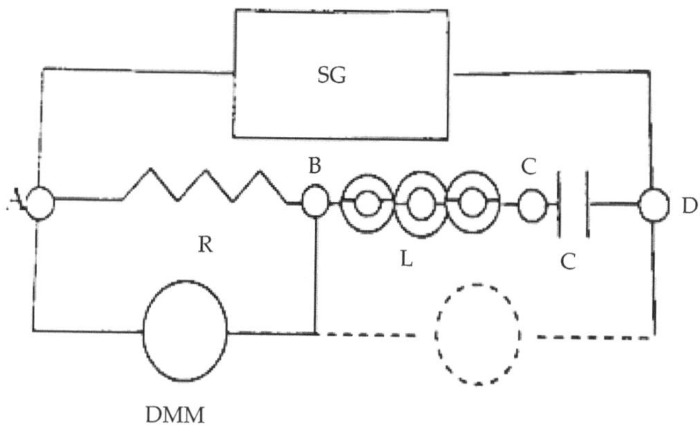

Fig. 1 : Circuit diagram for Series Resonant Circuit.

Connect the signal generator output to the two terminals A and D. Set the frequency of the signal generator at 1 kHz. Set the Root Mean Square (RMS) amplitude of the output at around 1 Volt. Measure the voltage across A and B terminals with a multi-meter in the AC 2 Volts range. This gives the voltage VR. With the same multi-meter measure, the voltage across the terminals B and D. This voltage Vz is the voltage across the series combination of L and C. Repeat the experiment at different frequencies from 1 to 7 kHz. The resonance frequency of the above coil and capacitor has a value of about 2.68 kHz. So in the range 2 to 4 kHz take readings at intervals of 200 Hz. Plot the graph of Vz/Vr against the frequency f. This ratio is $|Z|/R$, where Z is the impedance of the series L and C combination. The graph will show minimum impedance at a frequency fmin. Compare this with the value of $f_{res} = (1/2\pi)(1/LC)^{0.5}$.

A plot of V_Z/V_R against the frequency in kHz is shown in Figure 2. The magnitude of the impedance of the coil and capacitance in series is $(V_Z/V_R)R$ and since R is 1 k-Ohm, (V_Z/V_R) gives the impedance of the series resonant circuit in k-ohms. From the figure we see that the impedance goes through a minimum between 2.6 and 2.8 kHz. This should be compared with the resonance frequency of the coil and capacitance that is 2.68 kHz.

Observation

(i) For Series Resonant Circuit

TABLE 1

Series Resonant Circuit

f in kHz	V_R in volts	V_Z in volts	V_Z/V_R	V_{app} in volts

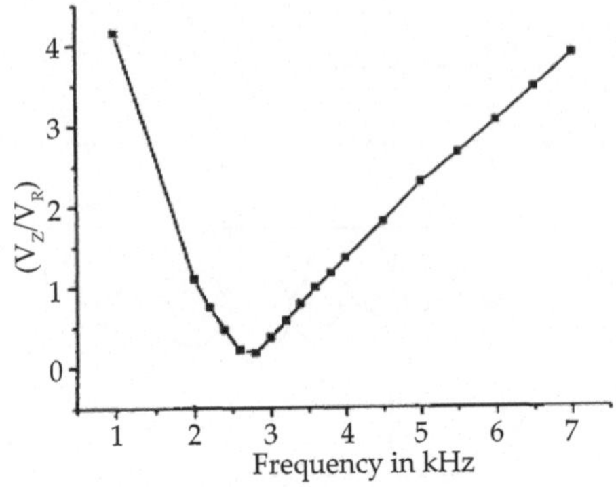

Fig.2 : A plot of VZ/VR against frequency in kHz for the series resonant circuit.

Next connect the terminals B and D together by an external wire and connect the signal generator between A and C. This is shown below.

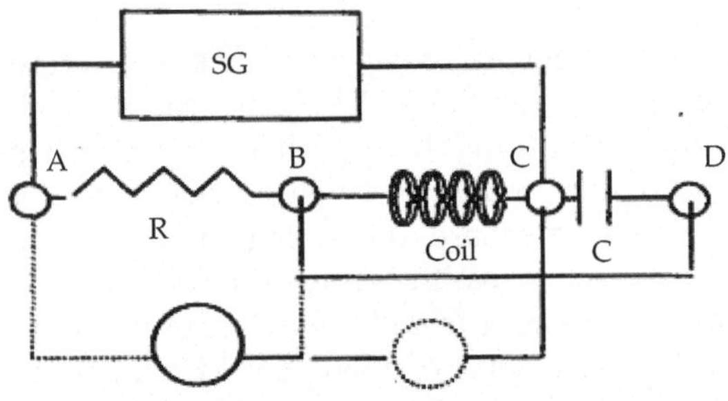

Fig. 3 : Circuit diagram for the Parallel resonant circuit.

As before, set the signal generator frequency at 1 kHz and the RMS output at around 1 Volt. Measure the voltage between A and B on a multi-meter in the AC 2 Volts range. This is V_R the voltage across the resistance R. With the same multi-meter measure the voltage V_z between B and C. This is the voltage across the impedance due to L and C connected in parallel. Vary the frequency from 1 to 7 kHz. From 2 to 4 kHz vary it in intervals of 200 Hz. The graph of V_z/V_R shows a sharp maximum at a frequency fmax. Verify that fmax is nearly equal to fres=$(1/2\pi)(1/LC)^{0.5}$.

Since R is 1 k-Ohm, (V_z/V_R) is the impedance of the parallel resonant circuit in kilo-ohms. Figure 4 shows the plot of (V_Z/V_R) against the frequency in kilo-Hz for the parallel resonant circuit. We see that the impedance of the parallel resonant circuit goes through a maximum at a frequency around 2.8 kHz again close to the resonant frequency of 2.68 kHz. The peak is much sharper than the minimum for a series resonant circuit.

(ii) For Parallel Resonant Circuit

TABLE 2

Parallel Resonant Circuit

f in kHz	V_R in volts	V_Z in volts	V_Z/V_R	V_{app} in volts

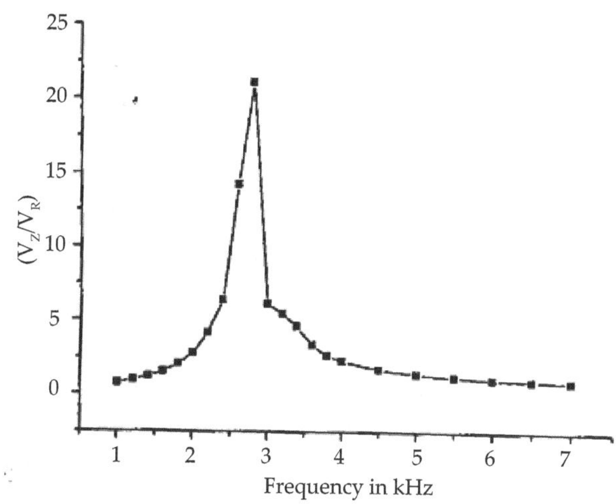

Fig. 4 : A plot of V_Z/V_R against frequency in kHz for the Parallel resonant circuit.

Result

The condition that the impedance of a circuit, in which an inductance L and a capacitance Care in series, is a minimum at the resonance frequency f_{res}, and the impedance of a circuit, in which L and C are in parallel, is a maximum at f_{res} have been proved.

Precautions

(i) With the help of signal generator, set the frequency at 1 KHz.

(ii) The graph must be drawn smoothly by choosing a convenient scale for the values.

Viva Voce

1. What is resonance of a circuit?
2. Explain series and parallel resonance?
3. What type of signal, the signal generator produces in your experiment?
4. What are the applications of series and parallel resonant circuits?
5. What are tuned circuits? Explain.

Experiment No. 10

Maxwell's Bridge

Aim

To compare the self-inductances of two coils L1 and L3 and find the balancing condition of Maxwell's Bridge.

These coils are connected as shown in the figure below

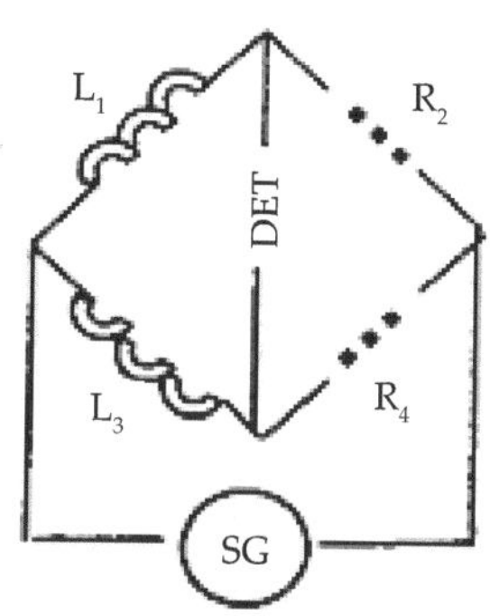

Fig. 1 : Maxwell,s Bridge Balancing condition is $L_1/L_3 = R_2/R_4$

Apparatus Required

Signal generator, AC Bridge set up, a DMM to measure AC volts to three decimal places and a DMM to measure resistance up to 2 k-ohms.

Description

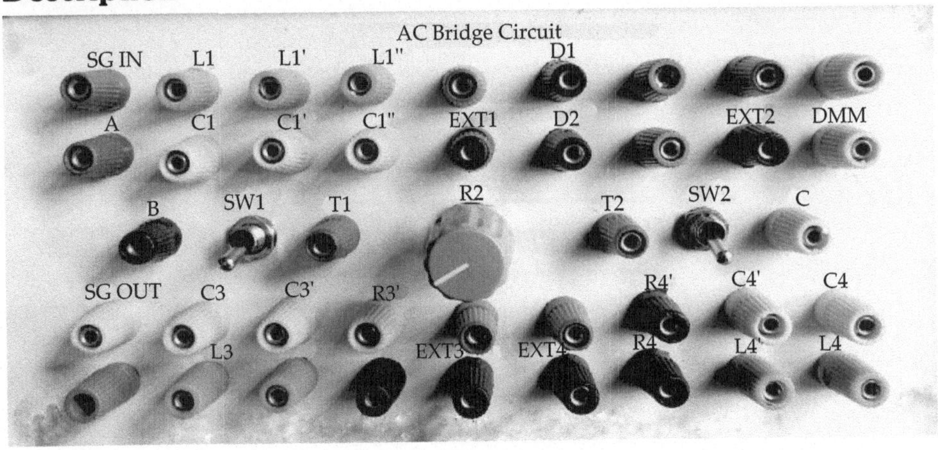

Fig. 2 : AC Bridge Circuit

There are three AC Bridge circuits, which can be realized with the above instrument. These are the Maxwell's bridge, the De-Sauty's bridge and the Maxwell-Wien bridge.

An AC bridge is a Wheatstone's bridge with impedances connected in the four arms as shown in Figure 1. The figure is drawn to accord with the symbols on the AC Bridge circuit. SGin and SGout are the two terminals to which the signal generator is to be connected. The terminals marked A and B are shorted with SGin while the terminal C is shorted with SGout. Between A and D1 is the impedance Z1. Between D1 and D2 is the DMM in the AC 200 mV range. In the front panel of the bridge D1 and D2 are connected to two terminals marked DMM. The DMM is to be connected between these two terminals. One can connect any impedance between A and D1. For example one terminal of a coil is connected to D1 internally and the other end of the coil is connected to a free terminal marked L1. If the terminal A is connected to the terminal marked L1 the coil takes the place of Z1. In this manner by connecting A to the terminals marked L1, L1′, L1" three different inductances can be connected as Z1.

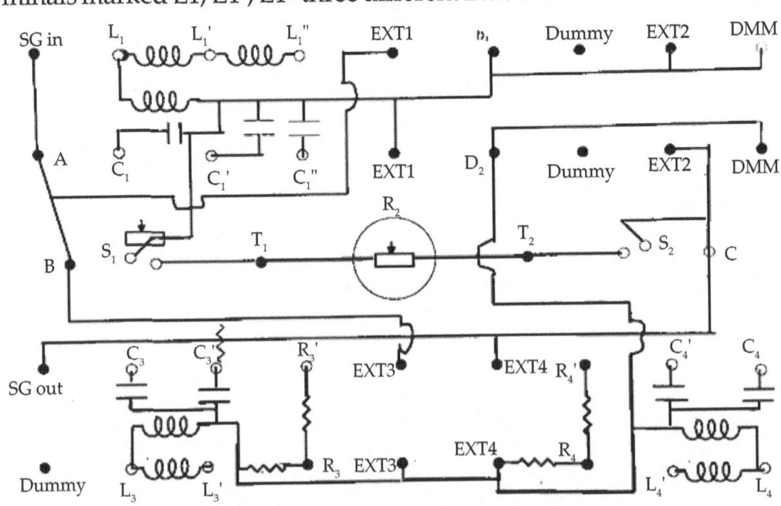

Fig. 3 : AC Bridge Circuit diagram

Similarly by connecting A to the terminals marked C1, C1' or C1" one may connect three different capacitances as Z1. In the third arm of the bridge circuit one end of the impedance Z3 is connected internally to D2. By connecting the free end of Z3 to the terminal marked B one may connect the impedance Z3 in the third arm of the bridge. In the third arm we may connect inductances L3 or L3', capacitances C3 or C3' and resistances R3 or R3′. Similarly by connecting the terminal marked C to L4, L4', C4, C4' or R4, R4' one may connect an inductance, capacitance or resistance as the impedance Z4. The arm Z2 is a variable resistor (pot). This pot is connected between D1 and C when the switches SW1 and SW2 are pressed down. When they are up the pot is isolated and its resistance can be measured by connecting a DMM between the terminals marked T1 and T2.

External terminals are also provided marked EXT1, EXT2, EXT3 and EXT4 for connecting external inductances, capacitances or resistances in place of the built in coils capacitances and resistances. In this case the terminals A, B and C are NOT connected to any other terminals and the switches SW1 and SW2 are in the up position. At the terminals marked EXT2 one connects a dial resistance box and balanceis obtained by adjusting the dials. In this way one may use impedances of one's choice.

TheMaxwell's Bridge can be realized by connecting A to L1 (L1' or L1"), B to L_3 (or L_3') and C to R_4 (or R_4′) one realizes the Maxwell's bridge.

Experimental Procedure

Connect the signal generator between the terminals marked SGin and SGout on the bridge. Adjust the frequency to be about 5 kHz and set an output voltage between 0.5 to 1 V. Connect the terminal A to L1, terminal B to L3 and terminal C to R4. Put the switches S1 and S2 on. Connect a multimeter in the AC milli-volts range to the terminals DMM on the bridge. Change the potentiometer R2 till the reading on the multimeter is a minimum. Then switch off S1 and S2 and measure the resistance R2 between the terminals T1 and T2 with a different multimeter. While balancing the bridge the multimeter to measure resistance should NOT be connected to T1 and T2.

Observation

Connect C to R′4and repeat the operations. Then connect B to L3' and repeat the observations. Then connect A successively to L1' and L1″ and repeat the observations. A sample set of readings are given below:

Frequency: 5 kHz   Amplitude: 0.5 V

L_3mH	R_4 Ohm	R_2 Ohm	L_1 mH	
				Coil L_1

				Coil L_1,

Result

The balancing condition of Maxwell's Bridge is L1/L3 = R2/R4.

NOTE : The coils sold in the market as having 3.6 mH inductance do not all have the same inductance. The inductances vary by as much as ±20%.

Precautions

(i) With the help of signal generator, set the frequency at 5 KHz.

(ii) While balancing the bridge, the multimeter to measure resistance should not be connected to T1 and T2.

Viva Voce

1. What is the working principle of Maxwell's bridge?
2. What is multimeter? Why it is named so?
3. What is signal generator? What type of signal is generated in it?
4. What is inductance and capacitance?
5. Name the physical quantities that can be measured using multimeter.

Experiment No 11

Refractive Index if the Material of The Prism

Object

To determine the refractive index of the material of the prism for the given sodium light.

Apparatus

A spectrometer, sodium lamp, a prism, reading lens and a lamp.

Description

Refraction through a prism - If we consider a ray of light PQ incident on the face AB of the prism ABC. Then at Q as it passes from a rarer into a denser medium, it is bent towards the normal and follows the path QR within the material of the prism. At R it passes from a denser into a rarer medium and consequently bends away from the normal along the direction RS. Originally the ray was travelling along the direction PQT but after passing through from the prism it takes up the path RS.

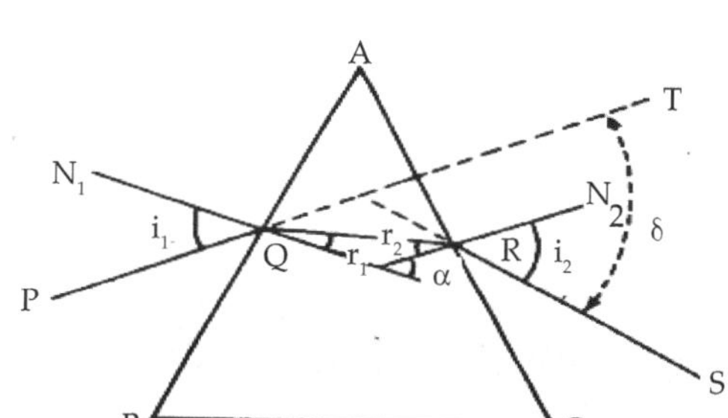

Fig. 1 Refraction through prism

From Fig. 1 it may be observed that the ray has been deviated from its original path by an angle δ. This angle is termed as the angle of deviation. From the geometry of the fig.1, we can deduce that,

$$a = r_1 + r_2$$

$$\text{and } \delta = (i_1 - r_1) + (i_2 - r_2)$$

$$= i_1 + i_2 - a$$

where the angles have their usual meaning.

It is clear that the base of the prism plays no role in producing the deviation of the ray, it is only the angle of incidence i1, the refracting angle a of the prism and its material, which are responsible for deviating the ray through an angle from its original path. However, for i2 the angle of emergence depends on angle of incidence i1.

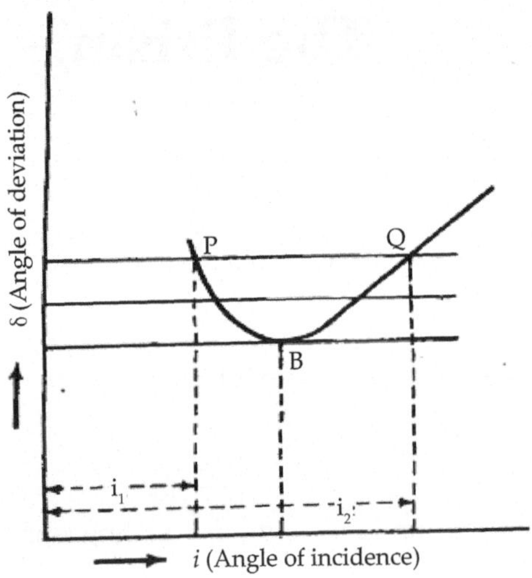

Fig. 2 Incidence deviation curve of a prism

It is interesting to examine the variation of deviation with the angle of incidence. Fig. 2 shows the incidence-deviation curve for a prism. It is also evident from fig.2 that the variation is not uniform. For small angles of incidence the deviation is large. As the value of i increases, the value of δ goes on decreasing. For a particular angle of incidence it attains minimum value. This angle of deviation is called the angle of minimum deviation. With further increase in the angle of incidence, the value of δ again increases. From the nature of the curve as shown in fig.2, a few important conclusions can at be drawn as stated below:

Considering PQ a line parallel to the axis representing the angles of incidence. PQ line intersects at the curve in two points P and Q. This means that for both the values of i (i = i1 and i=i2), the deviation produced by the prism is the same. It is also clearly understood from the figure 1. If the ray is incident along SR at an angle i2, it will emerge along QP at an angle i1 and the deviation suffered will be same as in the first case when the ray was incident along PQ. Then from the i –δ curve we can conclude that if i1 is the angle of incidence, i2 is the corresponding angle of emergence. Let PQ descend down towards the incidence axis but still parallel to it. The value of i1 increases whereas that of i2decreases. This continues till PQ becomes tangential to the curve at B. Here i1 = i2 ; the angle of incidence has become equal to the angle of emergence hence, it gives an important conclusion that a ray of light suffer minimum deviation when it passes through the prism symmetrically. The ray is not necessarily parallel to the base except in the case of an isosceles or an equilateral prism. When the ray passes symmetrically through the prism, it is said to be in the position of minimum deviation.

Here

$$a = r_1 + r_2$$
$$\delta = i_1 + i_2 - a = 2i = a$$

where,

$$i_1 = \frac{\delta + a}{2} \, and \, r = \frac{a}{2} (if \, r_1 = r_2)$$

The Spectrometer : Spectrometer is an instrument that produces the dipersion of an incident light which results in to the formation of the spectrum.

Three parts of Spectrometers are :

1. The collimator for producing an emergent beam of parallel light which falls on the prism.
2. The revolving prism table for mounting the prism used for producing the dispersion of the incident pencil of light.
3. The telescope for observing the spectrum. The description of each part in details are given below. Figures 3 and 4 describe the important parts of the spectrometer.

(1) **The collimator** — It consists of a main tube at one end of which there is a well-corrected convergent lens Lc and at the other end draw tube carrying a slit S. The draw tube can be moved inside the main tube by hand or by rack and pinion arrangement which results in to the fact that the slit S can be placed in the focal plane of Lc. The slit has one jaw fixed and the other jaw is movable by a screw. In some slits both the jaws may open out or close in simultaneously. The slit can be illuminated by the source of light the spectrum of which is to be studied. In general, the collimator is rigidly fixed. However, in some instruments it can rotate about the axis of the spectrometer.

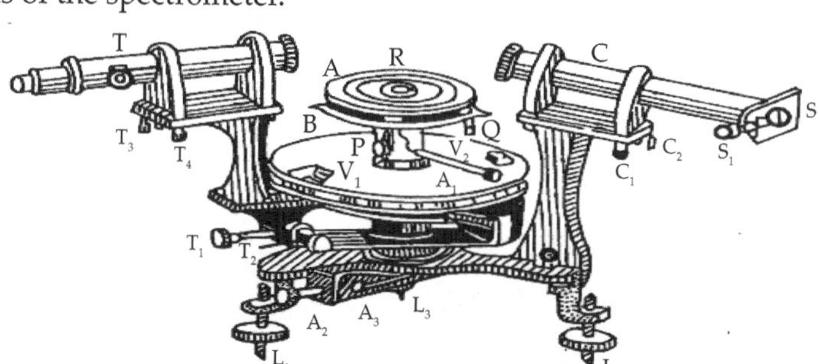

Fig. 3 Spectrometer

(2) The revolving prism table—The prism table P is a circular plane table which can be rotated up to 360oabout the axis of the instrument independently of the collimator C and the telescope T. This table carries two verniers moving over the main graduated circular scale LMN, which is being carried by the telescope. The height of the prism table can be adjusted by moving up or down the rod carrying the table and then clamping it in the desired position. The table is provided with three levelling screws, P.Q.R (Fig. 4).

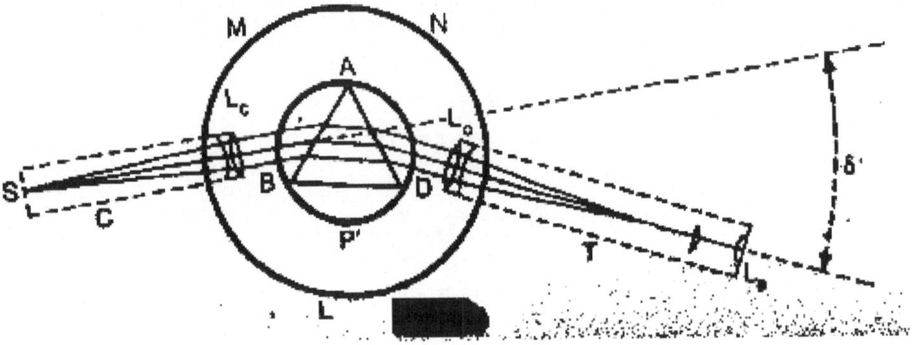

Fig. 4 Dispersion produced by prism

The concentric circles and straight lines parallel to the line joining two of the levelling screws are drawn on the surface of the table. This table is also provided with of clamp so that it can be fixed in any desired position. In the clamped position, slow rotation can be given with the help of a tangent screw at the base of the spectrometer.

The prism ABD can be mounted on the prism table such that it can receive the parallel beam of light coming from properly adjusted collimator. Dispersion is produced by the prism and from the face AD of the prism emerge various systems of parallel rays which can be brought to a focus by a telescope T.

(3) The telescope— The telescope T consists of a main tube carrying an objective L0 and a tube which can be worked out with rack and pinion arrangement and carries a Ramsden's eye-piece Le having a pair of cross-wire is formed. The telescope can be set so as to admit a parallel beam of light and an image upon the cross-wire which is distinctly visible through the eye-piece. The axis of rotation of the telescope coincides with the axis of the prism table. With its rotation, the telescope moves the main graduated circular scale. The number of degrees through which the telescope is rotated can be observed with verniers attached to the prism table. Like the latter it is also provided with a clamp and a tangent screw.

Theory

We have the relation,

$$\mu = \frac{\sin\frac{a+\delta}{2}}{\sin\frac{a}{2}} \qquad (1)$$

Where a is the refracting angle of the prism and δ is the angle of minimum deviation for the ray of the given wavelength. It is evident from equation 1 that the determination of the refractive index of the material of the prism for the given light involves the measurement of a and δ.

Procedure

1. The slit of the collimator is illuminated with sodium light. The is mounted prism on the table such that its centre coincides with the main axis of the instrument and the face AC lies normal to the collimator axis.
2. The angle α which has been measured should be used as the refracting angle. This method of mounting the prism not only corrects for want of parallelism in the incident beam of light on the face AC but also improves the resolving power of the prism since the entire beam from the collimator falls on the face AC.
3. Turn the prism towards the left and look for the dispersed image of the slit through the face AB by moving the eye in a horizontal plane containing the collimator axis. Having obtained the image, turn the telescope to point in that direction to receive it in the centre of the field of view.
4. Turn the prism further and follow the image in the telescope. Since in the initial setting the face AC was almost normal to the incident pencil it means that the angle of incidence on the prism was small. The rotation of the prism table towards the left results in to an increase in the angle of incidence which in turn results in to the decrease of the angle of deviation.
5. To follow up the dispersed image of the telescope, the prism should be moved towards the direction of the incident rays. As the prism table is turned further a position is reached when the rays suffer minimum deviation and the telescope is nearest to the line of the collimator axis so produced.
6. With further rotation of the prism table, the angle of deviation increases and the image begins to move backward from the position of the previous adjustment. Hence the position of minimum deviation can be obtained by adjusting the telescope in such a position that the dispersed image moves up to the cross wires and then begins to retrace its path with the rotation of the prism. This position of the telescope gives the direction of the emergent rays from the prism. Reading of both the verniers should be recorded.
7. Now, remove the prism without disturbing the prism table, which should be securely clamped. Unclamp the telescope and turn it so that its axis may be in line with the collimator axis. In such a case the direct image of the slit will be visible through the telescope and it can be finally set on the cross-wires by using the tangent screw of the telescope.

8. After completing step 7, read the two verniers and record their values. This position of the telescope gives the direction of the incident rays. The angle of deviation is the angle between the direction of the incident and the emergent rays. To calculate the angle of deviation δ, take difference of the readings of the same vernier for the two positions of the telescope. Take the mean of the two results. It gives the value of δ.
9. As a next step, turn the prism table so that the light from the collimator now falls on the face AB and the dispersed image has to be seen through AC. Repeat the experiment as above and get the value of δ. Finally get the mean value of δ by taking the average of the results obtained with the light incident first on the face AC and then on the face AB.
10. Substitute the values a and δ in the formula as given in Eq. (1) and calculate refractive index μ.

Observations

Measurement of the angle of the prism

Value of one main scale division = ______degree/minute

Total number of divisions on the vernier scale = ______

Least Count = ______ sec.

[A] Measurement of the angle of minimum deviation. When Light incident on the face AC.

Table 1

Sl. No.	Vernier	Reflection from face AB			Reflection from face AC			Difference (V_1-V_2)	Mean value = 2a.
		Main scale reading	Vernier reading	Total Reading (V_1)	Main scale reading	Vernier reading	Total Reading (V_2)		
1									
2	V_1								
3									
1									
2	V_2								
3									

[B] Measurement of the angle of minimum deviation.

Light incident on the face AB.

Table 2

Sl. No.	Vernier	For dispersed image			For direct image			Difference (V1-V2)	Mean value = δ
		Main scale reading	Vernier reading	Total Reading (V1)	Main scale reading	Vernier reading	Total Reading (V2)		
	V1								
	V2								

Calculations

$$2a =$$

$$a =$$

$$\text{Mean Value of } \delta =$$

$$\mu = \frac{\sin\frac{a+\delta}{2}}{\sin\frac{a}{2}} =$$

Result

The refractive index of the material of the prism for sodium light =.........

Precautions

1. The mechanical adjustments of the spectrometer should be correct, i.e.
   (a) The axis of the collimator and the telescope must intersect the main axis of the spectrometer, the latter being defined as the common axis of rotation of the prism table and
   (b) The axis of the collimator and the telescope must be perpendicular to the main axis of the instrument.
2. The optical adjustments of the spectrometer must be made correctly so as to ensure that
   (a) The eye-piece is focused on the cross-wires,
   (b) The telescope is adjusted to receive a parallel beam of light and form an image on the cross-wire without parallax.
   (c) To ensure that the collimator is in alignment with the source. Illuminate the slit of the collimator by placing the slit close to the window of the source of light. Open the slit wide enough and look through

the collimating lens. The lens should be fully and symmetrically illuminated by the light coming out of the slit. If this be not the case, the spectrometer as a whole should be displaced to achieve the adjustment. This adjustment is called alignment.

3. The slit used should be narrow : its width should be diminished so far as permissible.
4. Set the prism properly to correct for want of parallelism in the beam of light coming from the collimator.
5. At no time while taking observations should both the table and the telescope be unclamped. While reading the movements of the prism table, the telescope must be initially clamped and vice versa.
6. Read both the verniers ; this eliminates the error due to non-coincidence of the center of the divided circle with the axis of rotation of the telescope or the table. Check that the difference between the readings of the two verniers is very nearly 180º.

Viva Voce

1. What is wavelength of sodium light?
2. What is refractive index of a material? Is it same for all materials?
3. What is collimator?
4. Explain the role of telescope in this experiment.
5. How does the light is produced in sodium lamp?

Experiment No. 12

Measurement of Low Resistance

Aim

To measure low resistance (resistance less than one Ohm) with the DC technique.

Apparatus Required

Signal generator, lock-in amplifier (LIA), two-channel oscilloscope, low resistance box, one DMM to measure AC frequency and one Digital Multimeter (DMM) to measure output of lock in amplifier to two decimal places in the DC 20 V range.

Principle

In this box a resistance of 4.7 k (1/4 W) is connected with the low resistance and a 100-Ohm (1/4 W) resistor. The free end of the 4.7 k resistor is connected to the red banana terminal, the free end of the low resistance to the middle terminal (Green). A 100 Ohm resistor is connected to the middle terminal (Green) and the end terminal (in the box this terminal is yellow). Potential leads across the low resistance come to the RCA socket marked signal on the left top of the box. The voltage across the 100-Ohm resistor serves as the reference signal. This reference signal can be taken between the middle and end banana terminals on the box or from the RCA socket marked reference at the right top of the box.

The low resistance is made of insulated thin copper wire (SW 42). One free end of the wire is soldered to the 4.7 k resistor and the other free end to the 100-Ohm resistor. Two potential leads are soldered on the copper wire between the current leads. The free ends of the potential leads are soldered to the terminals of the RCA socket marked Signal.

The 4.7 k resistor limits the current through the wire to a few hundred microamperes when the signal generator output is connected to the terminals of the box.

Figure 1: Low resistance box.

Procedure

1. The output terminals of the signal generator are connected to the two end banana terminals on the box. The reference signal, taken from the RCA socket marked reference is connected to the RCA socket marked EXT REF on the front panel of the Lock- in amplifier. The RCA socket marked SIGNAL on the box is connected to the banana terminals marked EXT SIG on the Lock-in amplifier front panel. Connect the socket marked OUTPUT on the front panel of the LIA to a DMM in the DC 20 V range. Put the switches SW1 and SW2 on the LIA to the positions marked EXT and the switch SW3 up.
2. Switch on the signal generator. Adjust its frequency at about 200 Hz and its amplitude about 4 V (VAC). Switch on the lock-in amplifier. A DC voltage will appear on the DMM which will reach a positive maximum value as the phase adjust pot is turned to the right extreme position. Note the value of this maximum DC voltage VDC. Change the signal generator output VAC in steps of 0.5 V down to 2 V and note the voltage VDC. Repeat the experiments at frequencies, 400, 600, 800 Hz.
3. At each frequency plot a graph between VDC and VAC. Fit the points to a straight line and get the slope dVDC/dVAC.

$$dVAC = RdIAC \tag{1}$$

R is the total resistance in the primary circuit (4.8 k) and dIAc is the change in the current through the low resistance when the signal generator voltage is changed by dVAc. The output DC signal VDC is proportional to the voltage Vr across the low resistance r. When the current through the low resistance is changed by dIAc,the voltage Vr changes by dVr where

$$dVr = rdIAC \tag{2}$$

and this causes a change in VDC by dVDC given by

$$dVDC = \mu dVr = \mu r\, dIAC = (\mu r/R)\, dVAC \tag{3}$$

So,

$$dVDC/dVAC = (\mu r/R) \tag{4}$$

4. To ensure that the contribution to Vr from any residual inductance of the low resistance is small, when we work at low frequencies and we find that the slope dVDc/dVAc is independent of the frequency.

Observation

Table 1

Low Resistance Measurement

Frequency	200 Hz		400 Hz		600 Hz		800 Hz	
	V_{AC} Volt	V_{DC} Volt	V_{AC} Volt	V_{DC} Volt	V_{AC} Volt	V_{DC} Volt	V_{AC} Volt	V_{DC} Volt

A plot of V_{DC} against V_{Ac} for 400 Hz is shown in Figure 2.

Table 2

Freq Hz	Slope

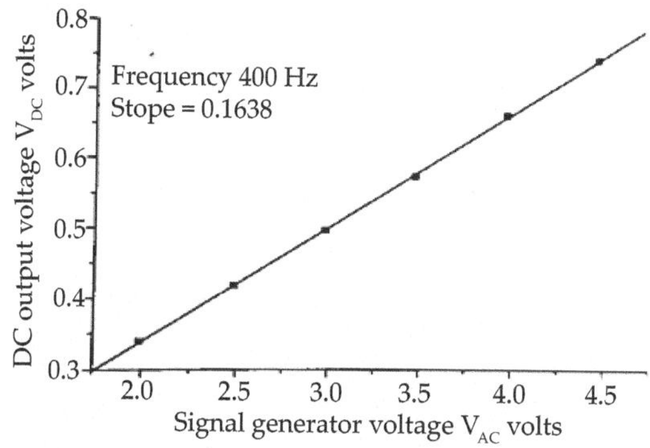

Figure 2 : A plot of V_{DC} against V_{AC} for 400 Hz

Result

From the linear fit the slope is found to be (______).

Precautions

(i) A DC voltage will appear on the DMM which will reach a positive maximum value as the phase adjust pot is turned to the right extreme position. The only note down the value of this maximum DC voltage VDC.

(ii) To ensure that the contribution to Vr from any residual inductance of the low resistance is small, always work at low frequencies.

Viva Voce

1. What is Ohm's law?
2. Why the experiment is designed to measure resistance less than 1 ohm?
3. Why slope of the curve is independent of frequency?
4. How your experiment is different than Ohm's law experiment to measure resistance?
5. What is the role of lock in amplifier in your experiment?
6. What are the applications of lock-in-amplifier?

Experiment No. 13

Transmission Characteristics of a Low Pass Filter

Aim

To study the transmission characteristics of a low pass filter.

Apparatus Required

Signal Generator, Filters box and a DMM measuring frequency up to 10 kHz and AC volts to three decimal places in the 2V range.

Introduction

A filter selectively transmits signals over a certain frequency range and blocks signals of other frequencies. The transmission of an ideal low pass is shown as a function of frequency in Figure 1.

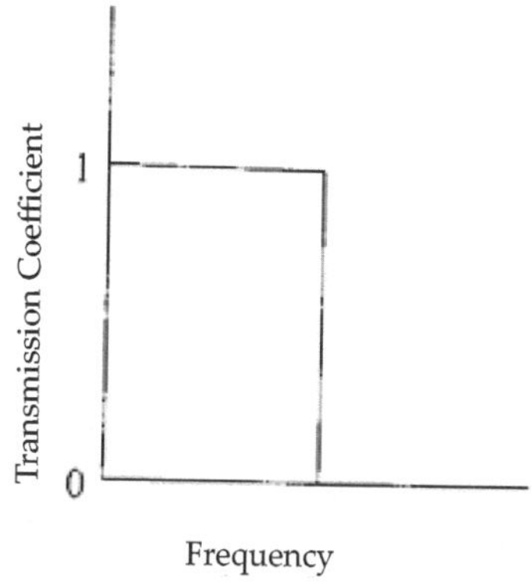

Fig. 1 : Transmission characteristics of an ideal low pass.

An ideal low pass filter will transmit all frequencies from zero to fmax without loss and blocks all the frequencies above fmax. An ideal high pass filter will block all frequencies below fmin and will transmit all frequencies above fmin without any loss. A band pass filter will transmit all frequencies between fmjn and fmax without any loss and blocks all other frequencies. An actual filter will not have such abrupt cut off transmission.

We shall consider passive filters as shown in Figure 2.

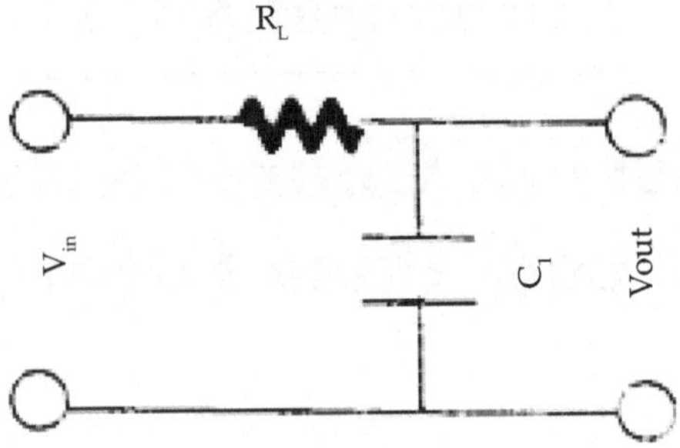

Fig. 2: Passive Filter (Low Pass)

The passive filters are constructed out of a resistance and capacitor in series. In the low pass filter one measures the output voltage across CL. At very low frequencies the impedance of the capacitance is much .higher than the resistance RL. Then almost the full input voltage Vin appears at the output terminals. When the frequency is very high, the impedance of the capacitance is very much smaller than the resistance. So the output voltage is very low. The transmission coefficient T defined as the ratio of the output signal to the input signal varies from nearly unity at very low frequencies to zero at very high frequencies. The roll off frequency (fRLP) for the low pass filter is the frequency at which the impedance of the capacitor is equal to the resistance. This happens at a frequency

$$fRLP = 1/(2\pi CLRL) \tag{1}$$

The roll off frequency increases as RL decreases. Since the voltage across the capacitance and the voltage across the resistance are $\pi/2$ out of phase, at the roll of frequency fRL,Vout is Vin/$\sqrt{2}$= 0.707 Vin. The transmission coefficient T = 0.707.

Principle

The low pass filters is built in a box. The top view of the box is shown in Figure 3. The low pass filter is on the left of the box. On the left are three banana terminals L1, L2 and L3. The low pass filter is connected between these three terminals, the resistance between L1 and L2 and the capacitance between L2 and L3. The resistance can be adjusted by the pot PLP. The knob of this pot is projecting to the left of the box. The pot PLP is set at around 150 Ohms. This is the value of RL. The capacitance CL between the L2 and L3 terminals has a value of 0.22 μf. This gives a value of fRLP of about 5 kHz.

Fig. 3: Top view of the filter box.

The two terminals L3 and H3 are connected together internally. If you find they are not connected by checking the continuity between them with a multimeter, connect a wire externally between the two terminals. The earth of the signal generator must be connected to one of these terminals.

Procedure

1. For studying the transmission characteristics of a low pass filter a signal generator is connected to the L1 and L3 terminals on the low pass side.
2. The frequency of the signal generator is set at 200 Hz and the RMS voltage Vinput is measured between the L1 and L3 terminals on the LP side with a DMM set in AC 2V range.
3. The output voltage Voutput of the low pass filter is measured between L2 and L3 terminals on the LP side with the same voltmeter. The frequency of the signal generator is changed in steps up to 10 kHz and the ratio of Voutput/Vinput is measured at each frequency.

Observation

Transmission coefficient of a low pass filter

RL~150 Ohms, CL = 0.22 μfd

Calculated Roll of frequency fRLP=4.82 kHz

TABLE 1

f kHz	log(f)	Vinput (Volts)	Voutput (Volts)	T	f kHz	log(f)	Vinput (Volts)	Voutput (Volts)	T

The Transmission coefficient is plotted as a function of frequency in figure 4.

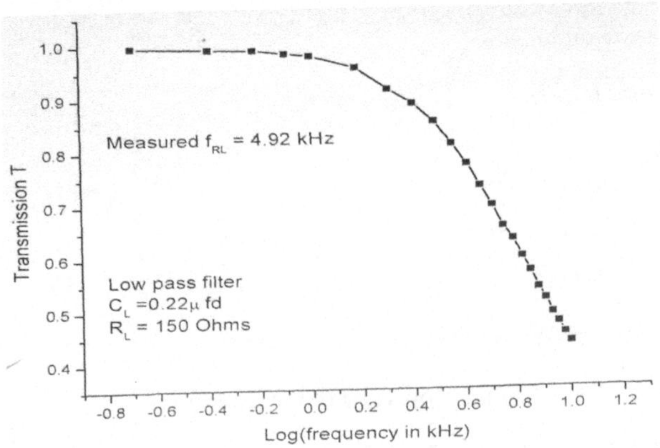

Fig. 4: Transmission coefficient as a function of frequency for the low pass filter.

Result

The log of the roll off frequency when the transmission is 0.707 is read off from the curve and is 0.6902. So fRLP = $10^{0.6902}$ = 4.92 kHz. This agrees well with the calculated value of 4.82 kHz.

Precautions

(i) The two terminals L3 and H3 are connected together internally. If you find they are not connected by checking the continuity between them with a multimeter, connect a wire externally between the two terminals. The earth of the signal generator must be connected to one of these terminals to bypass excess voltage.

(ii) The frequency of the signal generator must be changed in steps up to 10 kHz.

Viva Voce

1. What is a filter? Name various types of filters?
2. What is low pass and high pass filter?
3. What is working principle of a signal generator?
4. What is the application of a low pass filter?
5. What is transmission coefficient?
6. How oscilloscopes visualize the signal on the screen?
7. What are the uses of a CRO?

Experiment No. 14

Principle of Phase Sensitive Detection

Aim

Calibration of the lock-in amplifier.(Calibration means measuring the maximum output DC voltage of the lock-in amplifier for a small known AC signal voltage as a function of the frequency of the signal).

Apparatus Required

Signal generator, lock-in amplifier, two channel oscilloscope and a DMM to measure DC volts to three decimal places in the 2 V range.

Principle

Amplifiers are used to amplify a weak AC signal. However the amplifier also amplifies noise. Amplifiers have a bandwidth of several kilohertz. Usually noise is present over a range of frequencies while the signal is at a single frequency. If the signal is very weak, amplification alone will not enable us to pick out the signal against noise.

There are different sources of noise in an amplifier. Flicker noise is present in all electronic instruments and its power spectrum varies inversely with the frequency. This noise becomes a problem only if one works at very low frequencies. Then we have noise due to electromagnetic pick up from running motors, tube lights and so on. Such a noise has a peak in the power spectrum only at the frequency of the motor or the frequency of the mains. Such noise can be reduced by electromagnetic shielding. The third source of noise is thermal noise. This is of thermodynamic origin and cannot be avoided. Its power spectrum extends over all frequencies. So it is called white noise. If we have resistance R through which a current I is flowing the voltage across the resistance will vary randomly about the average value $V_o = IR$. The mean square fluctuation of the voltage is defined by $(V-V_o)^2$ where the average is taken over a long period of time. If we measure the noise over a bandwidth W, the mean square voltage will be

$$(V-V0)^2 = 4 \text{ kB T R W} \qquad (1)$$

To reduce this noise we either reduce the temperature T or reduce the bandwidth W of the amplifier. For detecting very weak signals at room temperature we have to effectively reduce the bandwidth to a few Hertz though the amplifier has a large bandwidth in kilohertz range. This is achieved by Phase Sensitive Detection.

Procedure

1. Connect a signal generator to the RCA socket marked SIG GEN at the bottom of the front panel of the LIA. The current drawn from the Signal generator will be very low. There is no need for a power amplifier.
2. Throw the switches SW1 and SW2 to the position marked CAL. Then the internal calibration circuit is connected to the LIA chip.
3. Set the frequency of the signal generator at 500 HZ. Adjust the amplitude Vapp of the signal generator at about 4 V.

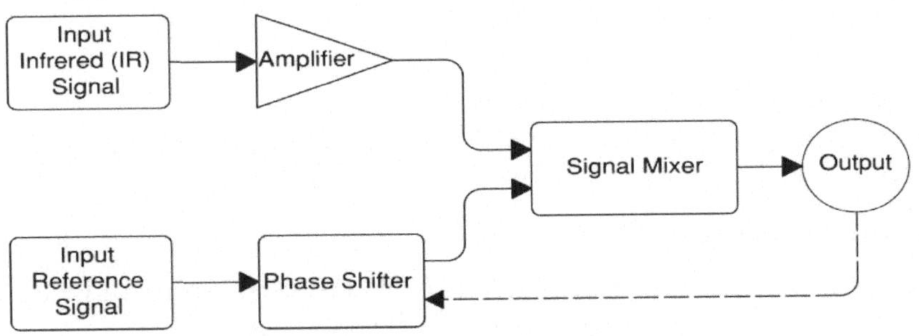

Fig. 1 Block Diagram Lock-in amplifier

4. Inside the Lock in amplifier, this voltage will be applied to three resistances in series, namely 220 k, 10 Ohms and 5 k. The voltage across 10 Ohms will be Vsignal.

$$\text{Vsignal} - \text{VappX } 10/ (225\text{x}103) = 44.4 \text{ Vapp}\mu\text{V} \qquad (2)$$

The reference voltage is the voltage across the 5 k resistor, one end of which is grounded.

5. Put the switch SW3 up. Turn the phase shifter potentiometer knob to the right extreme (resistance is zero and phase shift is zero).
6. Connect the RCA sockets marked REF and REF′ to the two channels of an oscilloscope. There will be two sinusoidal traces at the frequency of the signal generator.
7. These are amplified reference signals, one before phase shifting and the other after phase shifting. Since the phase shift is zero when the phase adjusting pot is at the extreme right, the two traces will be superposed exactly indicating zero phase shift.

8. Note the DC volts on a DMM in the 2 V range connected to the RCA socket marked OUTPUT. This DC voltage will be positive. Turn the phase shifter potentiometer knob anti-clockwise and observe what happens to the signal coming from REF′ relative to the signal coming from REF. You will see that the signal from REF′ moves to the right relative to the signal from REF. This indicates a phase shift between the two signals.
9. Note that as you turn the knob, the DC voltage on the multimeter comes down. Turn the knob and adjust to the position when the DC voltage on the multimeter is zero. You will see that the maxima of the REF′ signal now occur at the position of zeroes of the REF Signal. This indicates a phase shift of 90°. You can check this by feeding the two signals to the X and Y plates of the oscilloscope and noting that the Lissajous pattern is a circle on the oscilloscope. (There is distortion of the reference signal after phase shifting. So the circle will be distorted.)
10. If the knob is turned further, the phase shift increases beyond 90°, the panel meter reading becomes negative. When the knob of the phase shift potentiometer is to the left end the phase shift is nearly (but less than) 180° and the panel meter indicates the negative largest value. In the calibration circuit, the signal and reference voltages are derived from the potential differences across two resistances in series. They are in phase. So the panel meter voltage is largest when the phase shift is zero.
11. Connect the RCA socket marked PIN 13 to the oscilloscope and see how the output of the lock-in chip changes as the phase shifter knob is turned. As the knob is turned the output at PIN 13 develops a larger and larger negative part.
12. Turn the phase adjustment potentiometer knob to the right extreme and wait till the DMM connected to the output socket of the lock in shows a steady DC value. At the right extreme the phase difference between the signal and the reference is zero. Note the reading on the DMM. This is called VDC.
13. Repeat operations 8 for various values of the output of the signal generator from 1.5 V to 4V in steps of 0.5V, keeping the frequency fixed.
14. Plot Vdc vs. Vsignal calculated from vapp using equation (1). Find the slope μ. This is the amplification factor at the frequency f (500 Hz).
15. Change the frequency from 500 Hz to 1000, 1500, 2000, 2500, and 3000 Hz. For frequencies at and above 1000 Hz, put the switch S3 down. At each frequency repeat step (13).

Observation

TABLE - 1

f (Hz)	500			1000			1500	
VAC Volts	Vsig μV	VDC Volts	VAC Volts	Vsig μV	VDC Volts	VAC Volts	Vsig μV	VDC Volts

In converting VAC into Vsignal we use the conversion factor 10/225 x 10^3 = 44.4 μV for one Volt from the signal generator.

Figure shows the calibration curve for 500 Hz. From this figure we see that the Lock-in amplifier output in DC Volts is proportional to the signal. The linear fit has a slope of (5.35 ± 0.022) x 103. Similar fits are done at all the other frequencies and the slopes are collected from the Table I.

Calculation

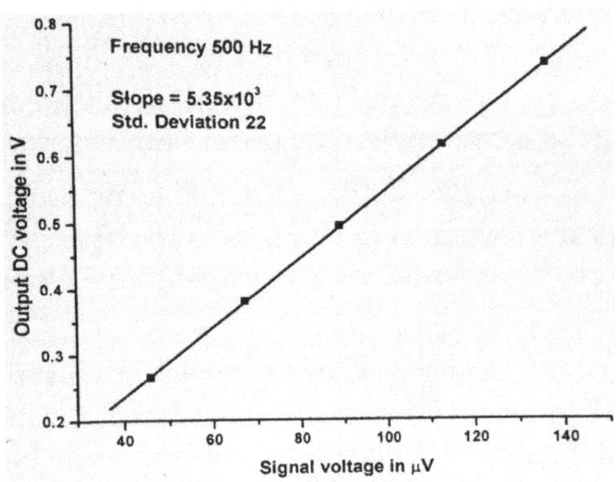

Figure 2 : VDC in V versus Vsignal in μV for 500 Hz

TABLE 2

fkHz	μ	x 103

Result

We see front the above table that the amplification factor can be taken to be a constant with a value .... x 103 independent of frequency.

Precautions

(i) For detecting very weak signals at room temperature we have to effectively reduce the bandwidth to a few Hertz, though the amplifier has a large bandwidth in kilohertz range. This is achieved by Phase Sensitive Detection.

(ii) While repeating step number 8 for various values of the output of the signal generator from 1.5V to 4V in steps of 0.5V, the value of frequency should be fixed.

Viva Voce

1. What is Flicker noise?
2. What do you mean by phase sensitive detection?
3. Why amplification factor is independent of frequency?
4. What is noise factor and noise figure?
5. What is the unit of noise?
6. What is the difference between noise and distortion?

Experiment No. 15

Vertical Torsion Apparatus

Object

(a) To verify that the angle of twist for a given length of a cylindrical rod or wire is proportional to the couple applied and (ii) the angle of twist produced by a given couple is proportional to the length of the rod or wire twisted.

(b) To calculate the value of modulus of rigidity for the material of the rod or wire (..........) – statical method.

Apparatus

Torsion Apparatus, Meter Scale, 0.5 kg weights, a vernier callipers and a screw gauge.

Description of Apparatus

This apparatus is due to Barton and is illustrated in Fig. 1 wire of the material to be tested is fixed at the top to a rigid support and at the bottom to a heavy cylinder C of brass. The cylinder being heavy keeps the wire vertical. Two parallel flexible threads leave opposite sides of the cylinder tangentially, and passing over two small frictionless pulleys A, B carry at the other ends pans of equal weights. Equal loads are placed in the two pans and the twists produced at three known distances from the fixed end of the wire are measured by means of double-end horizontal pointers P1, P2 and P3 which move over scales graduated in degrees.

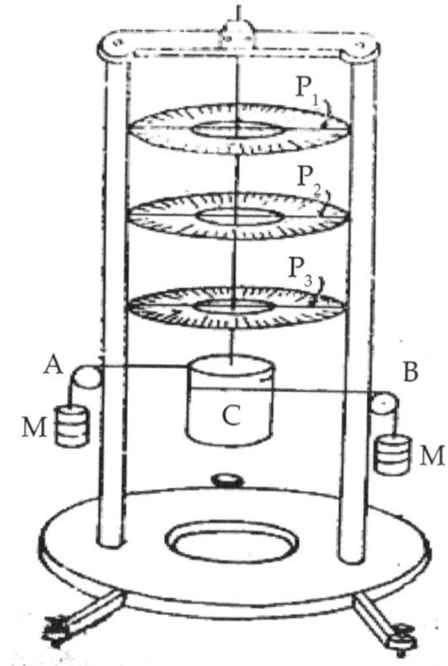

Fig. 1 Barton's Torsion Experimental Setup

Principle

If the load placed in each of the pans is m and the change in twist produced at a distance l from the fixed end of the wire of radius r is ϕ, then the couple due to equal loads M in the two pans is Mgd, where d is the diameter of the heavy brass cylinder ; and the couple due to ***torsional reaction*** is $\frac{\pi n r^4 \phi \pi}{2l180^o}$

Equating these two couples $$\frac{\pi n r^4 \phi}{360^o l} = Mgd$$

Whence $$n = \frac{360^o Mgd}{\pi^2 r^4 \phi}$$

The uncertainty in the exact position of the wire where it is clamped can be eliminated by noting down the twists $\phi1$ and $\phi2$ for two lengths l1 and l2 . Then evidently the twist for the length (l1 – l2) of the wire will be($\phi1$-$\phi2$) and hence the above expression for n become

$$n = \frac{360^o Mgd(l_1 - l_2)}{\pi^2 r^4 (\phi_1 - \phi_2)} \qquad (1)$$

This formula can be used to calculate n.

Procedure

1. First of all, level the base of the Barton's apparatus by using the levelling screws testing the levelling with a spirit level or a plumb line arrangement if the latter is incorporated in the apparatus. Adjust the pulleys A and B such

that they are at the same level and see that the strings leaving the cylinder in opposite directions are parallel and also tangential to the cylinder and the pulleys. Clamp the pointers (say at 40 cm, 60 cm, and 80 cm), from the fixed end of the wire adjusting their positions near about zero mark on the circular scales. Note down the zero readings of the pointers at their two ends, i.e., with no load in the pans. Take mean of the zero readings of the pointers at their two ends. The mean of the two readings at the ends of the same pointer will be free from any error due to the eccentricity of the axis of the wire with respect to the circular scale.

2. As a next step place gently a load of, say 20 gm., on each pan. This will change the value of the applied couple and hence the twist in the wire. Again note down the readings of the pointers at their two ends. The change in the reading of the same pointer evidently gives the twist in the wire by the additional couple applied for the length of the wire up to that pointer.
3. Now gradually increase the load on the pans till maximum permissible load is reached taking down the readings of the pointers at both ends after addition of each load.
4. Now, decrease the load on the pans in the same stages to zero noting down the readings of the pointers after the removal of each load. Take the mean of the readings of the same pointer for various loads taken with increasing and decreasing loads, if mean values agree closely, and then by subtracting the mean reading for zero load from the mean reading for any load, find the twists for various loads placed on the pans.
5. Now measure the lengths of the wire up to the pointers from its fixed end and altering the position of the pointers on the wire, take similar sets of observations for at least three more lengths of the wire.
6. After completing step 5, determine the diameter of the wire with a screw gauge and that of the cylinder with a vernier callipers.

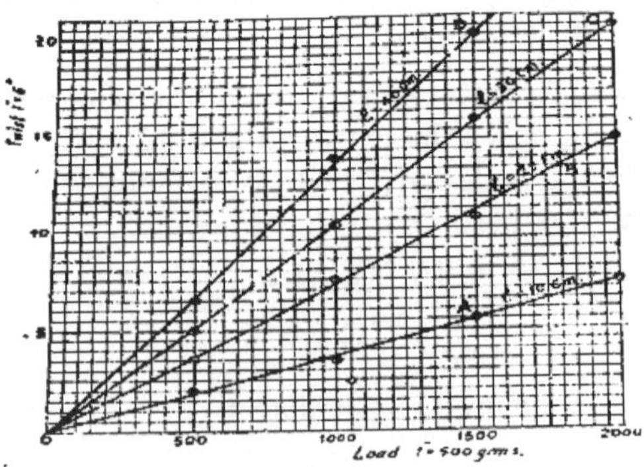

Fig. 2 Graph between load and twist

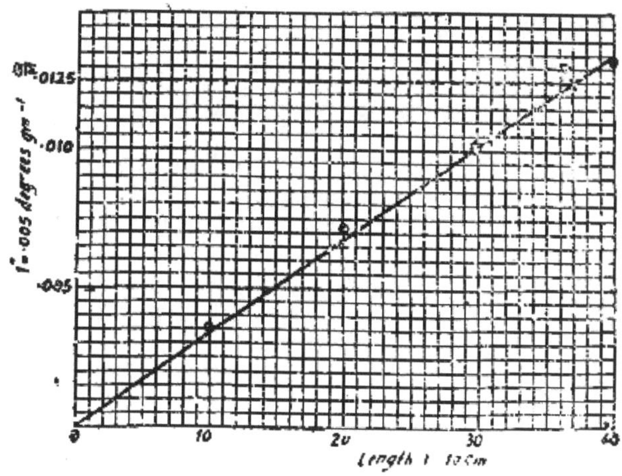

Fig. 3 Graph between ϕ/M and length l

7. Plot a graph between the load and the twist for each value of length of the wire as shown in fig. 2. The graph will be a straight line passing through the origin. From the graph find the best value of ϕ/M for each value of *l*.
8. Plot a graph as illustrated in fig.3, between ϕ/M and l. The graph will again come out to be a straight line. Take two convenient points on this line and find out the difference between their abscissae which gives *(l1 - l2)* and also the difference between their ordinates which gives ϕ1-ϕ2/*M* . Then calculate the value of modulus of rigidity from Eq. (1).

Observations

Diameter of the pulley :....................cm.

(A) Measurement of angle of twist ϕ

Length of the wire = ... cm

TABLE 1

Sl. No.	Lead M on the pan (gm)	Reading of the pointer in degrees: with increasing load				Reading of the pointer in degrees: with decreasing load			Mean Twist ϕ in degrees
		Reading at left end x	Reading at right end y	$\frac{x+y}{2}$	Twist	Reading at left end x′	Reading at right end y′	$\frac{x'+y'}{2}$	
1	0								
2	20								
3	40								
4	60								
5	80								
6	100								

(B) Diameter of the cylinder = ... cm.

(C) Measurement of diameter of the rod or wire :

(i) Least count of screw gauge = ... cm.

(ii) Zero Error = ... cm.

TABLE 2

Sl. No.	Reading along any diameter a (cm)	Length l of the rod or wire (cm)	$\frac{a+b}{2}$ (cm)	Mean diameter (cm)
1				
2				
3				
4				
5				
6				

Calculation

From the graph 1.

Point on the graph	Reading along perpendicular diameter b (cm)	**φ/M** *(degrees/gm)*
A		
B		
C		
D		

From the points P and Q on the graph (Fig. 2).

(l1 - l2) = ... cm.

(φ1 - φ2)/ M = ... degrees/gm.

Mean corrected radius of the wire = ... cm.

Mean radius of the cylinder = ... cm.

$$n = \frac{360^\circ Mgd(l_1 - l_2)}{\pi^2 r^4 (\phi_1 - \phi_2)} = \text{.........} \text{ dynes/cm}^2.$$

Result

The graph between φ and M for a given value of l and between φ/M and l are straight lines passing through the origin showing that the angle of twist is proportional to the value of the couple applied, and that if l is varied, the value of twist for a definite couple is proportional to the length of the rod or wire.

The value of modulus of rigidity for the material of the rod or wire (.......) = ...dynes/cm^2.

Standard Value = ... dynes/cm^2.

Error = ... %

Sources of Error and Precautions

1. The rod or wire should be clamped firmly other wise when loads are placed in the pan, it will turn a little in the clamps and the observations for angles of twist will be incorrect.
2. The vertical portion of the string should always be tangential to the pulley so that the force due to the load in the pan may always act along a tangent to the pulley.
3. For all metals Hooke's law holds only for small-hear and ceases to apply when the shear exceeds 1/200 radians or 1/3°, either because the metal breaks before this shear is reached (e.g., hard steel) or because it flows (e.g., lead). In practice the shear should not exceed 1/600 radians or 1/90. Hence angle of twist $\phi = l\theta / r < l / 9r$ degrees. When l= 50 cm., r=0.2cm., ϕ max. = 28º.
4. The masses should be placed or removed from the pan gently and the load in the pan should be increased gradually to a maximum in a number of equal steps and should be decreased in the same stages to zero.
5. As the radius of the rod or wire occurs raised to the fourth power in the expression for n and is a very small quantity, the diameter should be measured at a number of points along the length of the rod or wire and at each point two measurements should be taken along the two mutually perpendicular diameters.
6. The graphs should be smoothly drawn. The representative points should be distributed about the straight lines as evenly as possible and should lie as close to them as possible.
7. The base should be levelled so that the pillars and the pulleys are vertical.
8. The strings leaving the cylinder in opposite directions should be parallel and tangential to the cylinder.
9. Both ends of the pointer should be recorded. The mean, reading will be free from error due to eccentricity of the axis of the wire with respect to the circular scale.

For Vertical Torsion Apparatus Only

10. The base should be levelled so that the pillars and the pulleys are vertical.
11. The strings leaving the cylinder in opposite directions -should be parallel and tangential to the cylinder.

12. Both ends of the pointer should be recorded. The mean reading will be free from error due to eccentricity of the axis of the - wire with respect to the circular scale.

Viva Voce

1. What is Barton's apparatus?
2. Explain the angle of twist.
3. What is the modulus of rigidity?
4. What is the relation between the angle of twist and the couple applied?
5. What is the torsional reaction?

Experiment No. 16

Horizontal Torsion Apparatus

Aim

To calculate the value of modulus of rigidity for the material of the rod or wire using Torsion Apparatus.

Description of Apparatus

As illustrated in Fig. 1, one end of the rod to be tested is fixed to a block A and the other end is attached to a steel axle which is fixed to the pulley B. Both the bearing and the block A are firmly fixed to a stout base-board. The couple is applied to the rod by means of load placed in a pan suspended by a cord wound round the pulley B and the twists produced at two points at a known distance apart are measured by pointers P1 and P2 which can be clamped to the rod at any point and move over circular scales graduated in degrees.

Fig. 1

Principle

Let a load of mass M when placed in the pan change the angle of twist by ϕ radians at a distance l from the fixed end of the rod. Then, if R is the radius of the pulley over which the cord is wounds the couple acting on the rod is MgR. This couple is balanced by the couple of reaction due to elasticity of the rod whose value is $\pi n r^4\phi/2l$, where n is the modulus of rigidity of the material of the rod and r its radius.

Thus, $\pi n r^4\phi/2l = MgR$

Whence $$n = \frac{2Mgrl}{\pi r^4 \phi}$$

If the twist ϕ be measured in degrees, the expression for n is modified to

$$n = \frac{360MgRl}{\pi^2 r^4 \phi} \quad (1)$$

The value of n obtained from the above expression is subject to slight error on account of uncertainty of the exact position of the end of the rod inside the block A in which it is fixed. This error is, however, eliminated by determining the twists at two points of the rod at a known distance apart. If (l1+x) and (l2+x) are the actual lengths of the rod from its fixed end twisted through angles ϕ1 and ϕ2 respectively, we have from equation (1).

$$n = k\frac{(l_1 + l_2)}{\phi_1}$$

Where $$k = \frac{360MgRl}{\pi^2 r^4 \phi}$$

or $n\phi 1 = k(l1-x)$

Similarly, $n\phi 2 = k(l2-x)$

Subtracting we get, $n(\phi 1-\phi 1) = k(l1-l2)$

Therefore, $$n = k\frac{l_1 - l_2}{\phi_1 - \phi_2}$$

or $$n = \frac{360MgR(l_1 - l_2)}{\pi^2 r^4 (\phi_1 - \phi_2)} \quad (2)$$

from which n may be calculated.

Procedure

1. Suspend the pan on one side of the pulley by means of a chord wound on it and see that the vertical portion of the chord is tangential to the pulley. Clamp the pointers at, say 10 cm. and 20 cm, from the fixed end of the rod, adjusting their positions near about zeromark on the circular scales

of degrees and note down the readings of the pointers on the scales in this position. This gives zero readings of the pointers, i.e., when no load has been placed in the pan.

2. Smoothly, place a load of, say 500g in the pan which will cause the pulley to rotate till the couple due to torsional reaction of the rod balances the applied couple. The pointers will, therefore, be displaced from their Zero positions. Again note down the positions of the pointers on the scales. The difference between the two readings of the same pointer is evidently the angle of twist at the pointer for the couple due to the load in the pan.
3. Now, gradually increase the load in the pan in equal steps of 500g till the maximum permissible load is reached, taking down the readings of the pointers on the scales after each addition of load.
4. After step 3, decrease the load in the pan in the same stages to zero noting down the readings of the pointers on the scales after the removal of each load. The reading of the same pointer on the scale for a certain load in the pan taken with increasing and decreasing load agrees very closely. If they do not, the rod was not clamped sufficiently firmly between the clamps so that when the load was placed in the pan it turned a little in the clamp. In this case the readings will have to be rejected and a new set taken after having clamped the rod firmly.
5. From a satisfactory set of observations thus obtained, deduce the values of the angle of twist for various loads in the pan by subtracting the mean of the two readings for zero load from the mean of the two readings for any load.
6. Now, suspend the pan on the other side of the pulley and repeat the above observations. Find out the mean values of twist for various loads from these two sets of observations, for the same value of length of the rod. These values of twist will be free from any error due to slight wrong centering of the pulley on the axis of the rod.
7. In next step, measure the length of the rod from the fixed end up to the pointers and alter the positions of the pointers on it and take similar sets of observations for at least three other positions of the pointers say at 25 or 50 cm from the fixed end.
8. Next find out the pitch, the least count and the zero error of a screw gauge and with it measure the diameter of the rod. Now measure the diameter of the pulley with callipers and find out its mean radius R.
9. Now, plot a graph (Fig. 2)between the corresponding values of twist ϕ and load M for each value of the length of the rod, taking loads as abscissa and the angles of twist as coordinates. The graph between ϕ and M for each value of l will be a straight line passing through the origin. From the graph find out the best value of ϕ / M for each value of l.

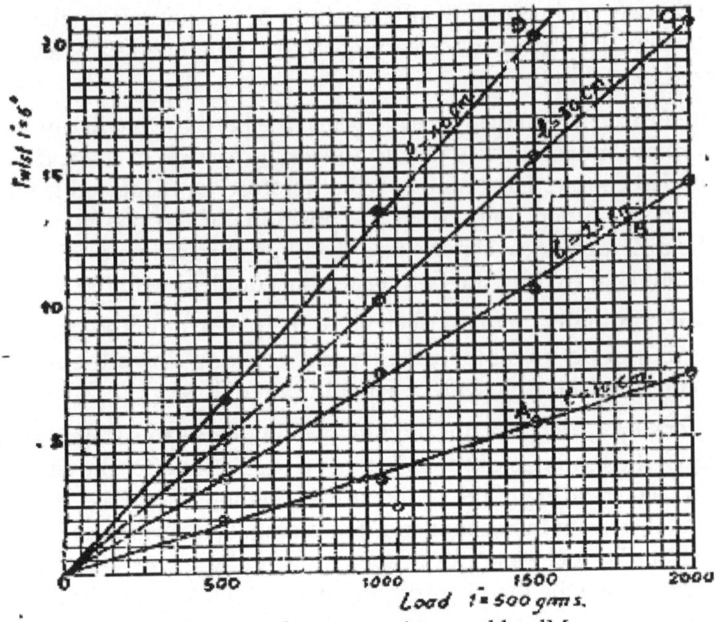

Fig. 2 Curve between twist ϕ and loadM

Next plot another graph (Fig.3) shown the corresponding values of ϕ/M and *l*. Here again the point should lie on a straight line passing through the origin. Select two points on this graph. The difference between their abscissa will give(l1 - l2) and the difference between their coordinates will give (ϕ1 - ϕ2)/ M which may be used to calculate the value of modulus of rigidity from the equation (2).

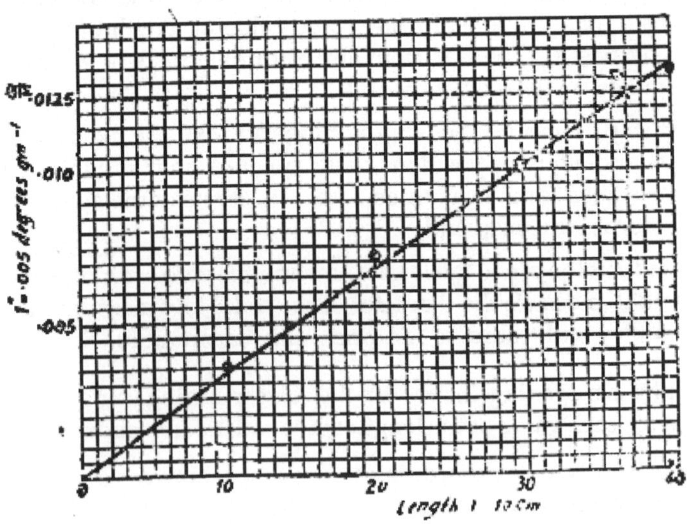

Fig. 3 Graph between ϕ/M and l.

Observations

(A) Measurement of angle of twist ϕ

Length of the rod = …… cm.

TABLE 1

Sl. No.	Load M on the pan gm	Cord to the left of pulley Reading of pointer in degrees				Cord to the right of pulley Reading of pointer in degrees				Mean twist ϕ in degrees
		with load increasing x	with load decreasing y	$\frac{x+y}{2}$	Twist	with load increasing x'	with load decreasing y'	$\frac{x'+y'}{2}$	Twist	
1	0									
2	500									
3	1000									
4	1500									
5	2000									
6	2500									

(B) Diameter of the pulley = …….. cm.

(C) Measurement of diameter of the rod or wire :

(i) Least count of screw gauge = ….. cm.

(ii) Zero Error = ….. cm.

TABLE 2

Sl. No.	Reading along any diameter (cm)	Length l of the rod or wire(cm)	$\frac{a+b}{2}$ (cm)	Mean diameter (cm)
1				
2				
3				
4				
5				
6				

Calculation

From the graph 1

Point on the graph	Reading along perpendicular diameter b (cm)	ϕ/M (degrees/gm)
A		
B		
C		
D		

From the points P and Q on the graph 2.

(l1 - l2) = ... cm.

(ϕ1-ϕ2)/M = ... degrees/gm.

Mean corrected radius of the rod = ... cm

Mean radius of the pulley = ... cm.

$$n = \frac{360MgR(l_1 - l_2)}{\pi^2 r^2 (\phi_1 - \phi_2)}$$ = ... dynes/cm^2.

Result

The graph between ϕ and M for a given value of l and between ϕ/M and l are straight lines passing through the origin showing that the angle of twist is proportional to the value of the couple applied, and that if l is varied, the value of twist for a definite couple is proportional to the length of the rod or wire.

The value of modulus of rigidity for the material of the rod or wire (.......) = ... dynes/cm^2.

Standard Value = ... dynes/cm^2.

Error = ... %

Sources of Error and Precautions

1. The rod or wire should be clamped firmly otherwise when loads are placed in the pan, it will turn a little in the clamps and the observations for angles of twist will be wrong.
2. The vertical portion of the string should always be tangential to the pulley so that the force due to the load in the pan may always act along a tangent to the pulley.
3. For all metals Hooke's law holds only for small shear and ceases to apply when the shear exceeds 1/200 radians or 1/3°, either because the metal breaks before this shear is reached (eg., hard steel) or because it flows (eg., lead). In practice the shear should not exceed 1/600 radians or 1/90. Hence angle of twist ϕ=lθ/r <l /9r degrees. When l= 50 cm., r=0.2 cm., ϕmax.=280.
4. The masses should be placed or removed from the pan gently and the load in the pan should be increased gradually to a maximum in a number of equal steps and should be decreased in the same stages to zero.
5. As the radius of the rod or wire occurs raised to the fourth power in the expression for n and is a very small quantity, the diameter should be measured at a number of points along the length of the rod or wire and at each point two measurements should be taken along the two mutually perpendicular diameters.
6. The graphs should be smoothly drawn. The representative points should be distributed about the straight lines as evenly as possible and should lie as close to them as possible.

7. The base should be levelled so that the pillars and the pulleys are vertical.
8. The strings leaving the cylinder in opposite directions should be parallel and tangential to the cylinder.
9. Both ends of the pointer should be read. The mean, reading will be free from error due to eccentricity of the axis of the wire with respect to the circular scale.

For Horizontal Torsion Apparatus Only

10. The zero positions of the pointers on the scales should be adjusted by placing the pan on the table with no load in it. In this case the zero positions of the pointers will be the same when the cord is on either side of the pulley.
11. To eliminate any small error of wrong centring of the pulley, observations for angles of twist should be taken by twisting the rod in both directions and taking the mean of the results.
12. If the string wound over the pulley is of appreciable thickness, half of its thickness should be added to the radius of the pulley to get the effective value of R.

Viva Voce

1. What is the difference between vertical Torsion apparatus with that of horizontal torsion apparatus?
2. What is Hooke's law?
3. Explain the terms stress, strain and shear.
4. Why are the precautions need to be taken to perform this experiment?

Experiment No. 17

Young's Modulus

Object

To determine the value of Young's modulus o.....from the flexure of a beam supported symmetrically on knife-edges near to its ends and loaded at its middle point.

Apparatus

A bar of the material whose Young's modulus is required, two knife-edges fixed in rigid supports in a horizontal plane, a stirrup carrying a hanger at its end, optical lever with the millimetre scale and the telescope (or a micrometer screw with electric bell, etc), a set of half-a -kilogram weights, a meter scale, vernier callipers, and a screw gauge.

Theory

Let a uniform rectangular bar supported symmetrically on two horizontal and parallel knife-edges be depressed by a load of mass m gm, suspended from the beam midway between the knife-edges. Then the depression δ of the middle of the bar is from eqn. (1) given by

$$\delta = \frac{mgl^3}{4bd^3Y} \tag{1}$$

where Y is Young's modulus for the material, l the length, b the breadth, add d the thickness (vertical) of the bar.

The depression δ being very small is measured with

(i) an optical lever or

(ii) a micrometer screw.

Method

Measure with a screw gauge the thickness of the experimental bar at several places along its length and find out the mean thickness d in cm, corrected for zero error of the instrument. Similarly take a number of readings for the breadth of the bar with vernier callipers and calculate the mean breadth in cm. Next place a meter scale on the two knife-edges with its graduated surface vertical, and with it measure the distance between the knife-edges. Then place horizontally on the knife-edges the experimental bar with its smallest edge vertical and with its equal lengths projecting beyond the knife-edges. The distance between the knife-edges already measured gives the length l of the bat between the knife-edges. Next suspend the hanger for placing weights from a stirrup resting on the bar midway between the knife-edges and also parallel to them.

Description

(a) Optical lever Arrangement

As illustrated in fig.1, place the optical lever on a fixed support behind the bar with its two hind legs on the support and with its front leg resting on the top of the stirrup at its middle point and adjust the face of the mirror to be vertical and also parallel to the length of the bar. Set up a vertical millimetre scale in front of the mirror at a distance of about one metre and close to the scale at the same height as the mirror, place a telescope firmly clamped in a rigid stand. The eye-piece is focused on the horizontal cross-wire of the telescope till it is distinctly seen. Then the telescope may be focused on the image of the scale in the mirror by turning round the mirror slightly, if necessary, about a horizontal axis.

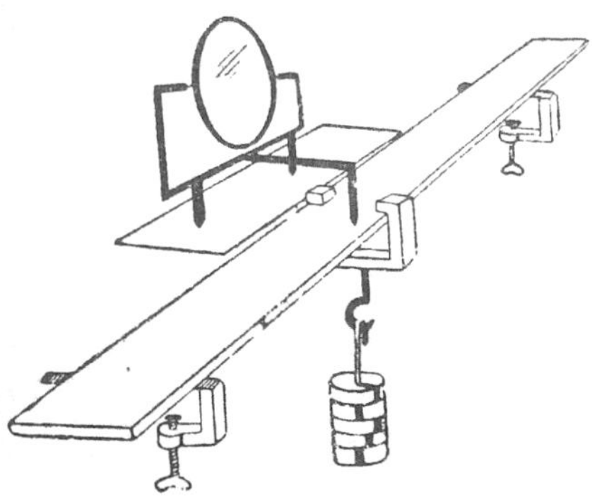

Fig. 1Experimental arrangement

In order to focus the telescope it is sometimes helpful to first focus it on the mirror itself and then by pushing the tube of the telescope in until the image of the scale appears in view. The position of the horizontal cross-wire of the telescope on the image of the scale is noted down.

As a next step, gently place a load of, say 500g on the hanger, which depresses the bar slightly and hence the mirror will be slightly tilted about the line joining its two back legs and the image of the scale in the mirror will be displaced past the cross-wire of the telescope. Again note down the position of the horizontal cross-wire of the telescope on the image of the scale is noted down. The difference between the two readings of the position of the horizontal cross-wire of the telescope on the image of the scale in the mirror clearly corresponds to the depression produced at the centre of the bar by the load placed on the hanger. Increase the load on the hanger by equal steps of 500g until a depression equal to nearly two-thirds of the maximum repression obtainable within elastic limit it reached, the Position of the horizontal cross-wire of the telescope on the image of the scale after each addition of load is noted down.

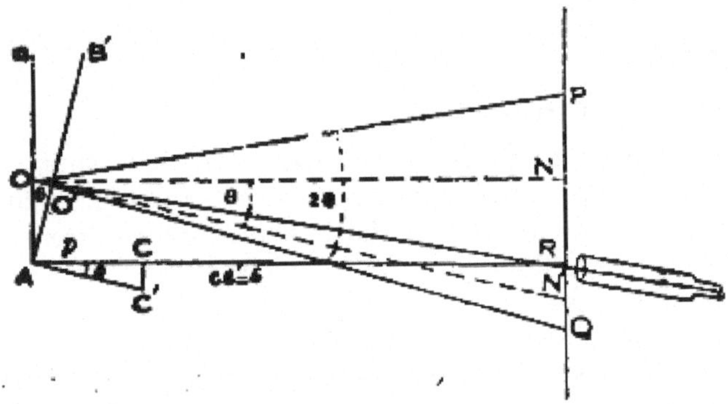

Fig. 2 Telescope Position for reading

Now as a next step, decrease the load on the hanger by the same stages to zero after noting down noting down the position of the cross-wire on the image of the scale in the mirror after the removal of each load. For various load, find out the mean of the two readings of the position of the cross-wire on the image of the scale in the mirror. Readings are taken with increasing and decreasing loads. From these means, the mean shift x of the image of the scale past the cross-wire of the telescope for a load is calculated.

Now, measure the distance D between the mirror and the scale. To measure p, i.e., the lever arm, place the optical lever on a sheet of paper and press it lightly such that it produces impressions of its feet. From these impressions the perpendicular distance of the front foot of the optical lever from the line joining the two back legs are determined, which gives p. Now the value of δ may be calculated from the formula

$$\delta = xp/2D$$

Finally, the value of Y is calculated from equation (1).

(b) Micrometer Screw Arrangement

Determine the pitch end least count of the micrometer screw and then set it up as described in Fig. 2, by making the connections for the electric bell as shown by the dotted lines. Now,the micrometer screw is lowered and micrometer reading is noted down when its tip just touches the bar as revealed by ringing of the bell. A load of, say 500g is placed gently on the hanger. This depresses the bar slightly. Lower the micrometer screw till it is again just in contact with the bar and note down its reading. The difference between the two readings of the micrometer clearly gives the depression produced at the centre of the bar by the load placed on the hanger. Now, increase the load on the hanger by equal steps of 500g until maximum permissible load is reached taking the reading of the micrometer screw after addition of load in each step.

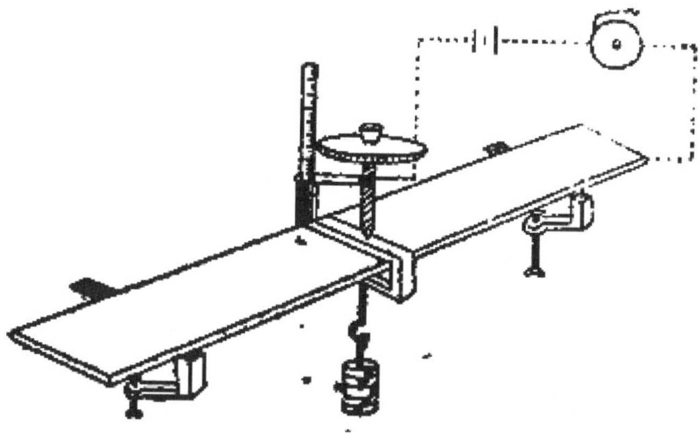

Fig. 3 Micrometer screw arrangement for reading.

As next step, keep the load on the hanger unchanged, move down the screw a little further and then move it up by turning it back till the tip of the screw is just not in contact with the bar. This will be indicated by just stopping of the ringing of the bell. The reading of the micrometer in this position will be the same as, or nearly equal to, its previous reading depending upon the amount of back-lash present in the screw.

Now, unload the hanger in the same stages till the load on the hanger is reduced to zero, by taking down the reading of the micrometer screw after removal of each load. Calculate the mean micrometer reading for various loads from the corresponding readings taken with increasing and decreasing loads and then calculate the mean depression δ of the bar for a particular load. Finally calculate the value of Y from equation (1).

Observations

(A) (i) Length of the bar between the knife edges = ... cm.

(ii) Measurement of thickness (vertical) and breadth of the bar

Least count of the screw gauge = ... cm.

Its zero error = ... cm.

Vernier Constant = ... cm.

Its zero error = ... cm.

TABLE 1

Sl. no.	Thickness cm	Breath cm
1		
2		
3		
Mean		

(B) Measurement of depression of the bar :

For Optical Lever Only

(i) Lever arm, ie, the distance p of the front foot of the optical lever from the line joining the other two = ... cm.

(ii) Distance D of the scale from the mirror = ... cm.

For Micrometer Screw Only

Pitch of the srew = ... cm.

No. of divisions on circular head = ... cm.

Least count of the micrometer = ... cm.

TABLE 2

Sl. No.	Load on the pan gm	Micrometer screw reading (or position of cross-wire on image scale)			Depression for 1500 gm	Mean
		Load increasing x (cm)	Load decreasing y (cm)	Mean $\frac{x+y}{2}$ (cm)		
1	0			a		
2	500			b	p(a,d)	
3	1000			c	q(b,e)	
4	1500			d	r(c,f)	
5	2000			e		
6	2500			f		

Calculations

For optical lever only

Multiplying factor of the optical lever, P/ 2D =

$$\delta = \frac{xp}{2D} = \text{.......} \text{ cm.}$$

Now, $$Y = \frac{Mgl^2}{4bd^3\delta} = \ldots\ldots\ldots \text{ dynes/cm}^2.$$

Result

The value of Young's Modulus for the material of the bar .... is .... dynes /cm^2.

Standard Value = ... dynes/cm^2.

Error = ... %

Sources of Error and Precautions

1. Since in the case of metals, Hooke's law fails if the elongation exceeds 1/1000 cm., per cm., the highest strained filament of the bar i.e. the outermost one should not be strained beyond this limit. For l = 80 cm., d =0.5 cm., maximum value of δ = 1.06 cm.
2. To secure symmetrical loading of the bar, the hanger used to carry weights should be suspended from a stirrup of the shape as shown in figs. 1 and 3, and the knife-edge of the stirrup should be parallel to the two knife-edges supporting the bar.
3. The masses should be placed on or removed from the hanger as gently as possible and the load on the bar increased or decreased gradually in a number of equal steps of, say 500g.
4. While calculating the mean depression from the mean micrometer (or optical lever) readings for increasing and decreasing loads, the student should be careful for not taking the successive differences between them.
5. As the thickness (vertical) of the bar occurs raised to the third power in the expression for Y and is also a small quantity, many observations for it should be taken at a number of points along the length of the bar with a screw gauge.

Viva Voce

1. Explain the term Young's modulus.
2. How does a telescope work?
3. Explain least count of a micrometer screw.
4. What is back-lash error?
5. What do you mean by a lever?
6. What is the relation between the different elastic constants?

Experiment No. 18

Coefficient of Viscosity of Water

Object

To determine the coefficient of viscosity of water by measuring the rate of its flow through a capillary tube of uniform cross section under constant pressure difference.

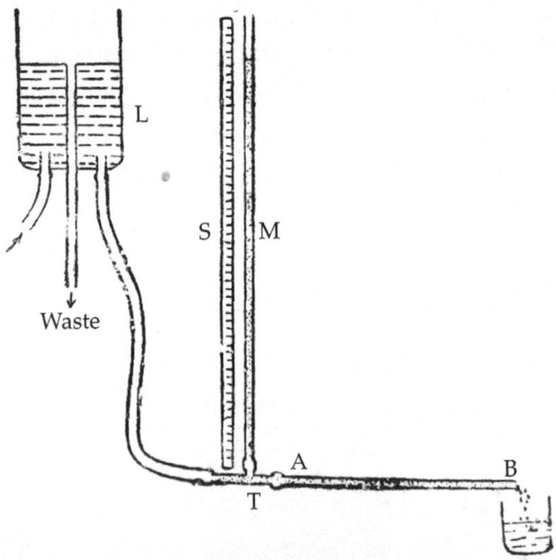

Fig. 1 Viscosity apparatus

Apparatus

Viscosity apparatus , a capillary tube, a beaker, a scale marked cylinder, a constant level water tank. a stop-watch, a microscope and a thermometer.

Description of Apparatus

As illustrated in Fig. 1, the apparatus consists of a capillary tube through which water is allowed to flow from a constant level water tank. The tank consists of a brass cylinder of about 7.5 cm. length and of about 7.5 cm internal diameter and is provided with inlet and outlet tubes near the sides and with a constant level over flow tube in the centre. The ends of the tubes are folded in to a series of small parallel folds for rubber tubing connections. The height of the tank from the capillary tube can be varied by sliding it along the supporting rod and can be clamped in any desired position with screw clamps. The end of the outlet tube of the tank is connected through an inverted T piece to the capillary tube, the tail of T being connected to a vertical glass tube which is used to measure the head of water.

Theory

When water is allowed to flow slowly through a uniform capillary tube of length l and of radius a, its volume flowing per second is given by,

$$Q = \frac{\pi p a^4}{8\eta l}$$

Where p is the pressure difference between the ends of the capillary tube and η the coefficient of viscosity of water.

Now, if h be the height of the free surface of water in the constant level water tank above the horizontal plane in which the axis of the capillary tube lies, as registered by the vertical glass tube, p = Pgh

where ρ is the density of water, and g is the acceleration due to gravity.

Thus,

$$\eta = \frac{\pi a^4 \rho g h}{8Ql} \qquad \text{.........(1)}$$

From this formula the value of η can be calculated if the rate of flow of water Q through the capillary tube is observed under a known constant head h.

Method

1. To begin with experiment, take a capillary tube of about 30-40 cm. length and of uniform cross section of about 0.5 mm in diameter and connect it horizontally at its one end through an inverted T-piece as shown in Fig. 1 to the outlet tube of the constant level water tank.
2. The water from a tap is allowed to run into the tank and its flow is adjusted at such a rate through the inlet tube of the tank that the level of the water in the tank is maintained constant. Then the flow of water is observed through the capillary tube is adjusted by altering the height of the tank

from the capillary tube till the liquid leaves the tube in a trickle. Collect sufficient quantity of water trickling from the end of the capillary tube in a scale marked jar during a carefully observed interval of time and find out its volume.

3. From this calculate the value of Q, the volume of water flowing per second through the capillary tube. Note down the head of water b, i.e., the height of the level of water in the constant level water tank from the capillary tube as registered by the vertical glass tube.
4. Now alter the pressure difference by moving up or down the water tank. Repeat the experiment with various values for b and calculate a mean value for h/Q. Also, note down the temperature of water.
5. As a next step the length l of the capillary tube is measured. The the capillary tube is mounted in horizontal position and a vernier or micrometer microscope is focused. The cross-wire tangential to the circular section is set on one side of its centre and the reading of the microscope is noted down. Then the cross-wire is set tangential to the other side of the centre of the section and again the reading of the microscope is noted down.
6. The difference between the readings of the microscope for the two positions of the cross-wire gives the diameter of the tube. Then the tube is rotated through a right angle and again the diameter of the tube is found out. Now, the microscope is focused on the other end of the tube and find out the two perpendicular diameters as above. The mean of these four readings for diameter gives the average diameter 2a of the capillary tube.
7. After obtaining a mean value for h/Q and after determing the values of l and a, the coefficient of viscocity η may be calculated by using equation (1).

Observations

[A] Measurement of h and Q.

Leastcount of stopwatch =........ sec.

Leastcount of graduated cylinder = ... cc.

TABLE 1

Set No.	h cm	Determination of Q				h/Q	Mean
		Volume of water collected c.c.	Time taken sec.	Volume flowing per sec. Q in c.c	Mean value of Q in c.c		
1							

[B] (i) Length of the capillary tube = ...cm.

(ii) Measurement of the diameter of the capillary tube Least count of the microscope = ...cm.

TABLE 2

Serial No.	Observations for position of cross-wire on the section of the tube											
	At one end of the tube						At the other end of the tube					
	Reading along any diameter			Reading along perp. diameter			Reading along any diameter			Reading along perp. Diameter		
	One end of section (cm)	Other end of section (cm)	Diameter (cm)	One end of section (cm)	Other end of section (cm)	Diameter (cm)	One end of section (cm)	Other end of section (cm)	Diameter (cm)	One end of section (cm)	Other end of section (cm)	Diameter (cm)
	Mean			Mean			Mean			Mean		

Diameter of the tube = ... cm.

[C] Temperature of water = ... ºC

Calculations

Mean Radius of the tube = Mean diameter of the tube/2 = ... cm

$$= \frac{\pi a^4 pgh}{8Ql}$$

= .... dynes per cm^2 per unit velocity gradient.

Result

The coefficient of viscosity of water at °C.

= .... dynes per cm.2 per unit velocity gradient.

Standard value = ...poise

Error = ... %

Sources of Error and Precautions

1. The cross section of the capillary tube should be about 0.5 mm., for the Poiseuille's formula holds good in the case of narrow tubes only. With wider tubes the velocity of the liquid exceeds the critical velocity and the flow becomes turbulent.

2. The capillary tube should be as uniform as possible in section- Its uniformity can be tested, if desired, by measuring the length of a short thread of mercury in various positions in the tube.
3. The capillary tube should be placed horizontally.
4. Since the radius occurs raised to the fourth power in the Poiseuille's formula, the diameter of the tube should be very accurately measured. Observations for diameter should be taken at both ends of the tube and at each end readings should be taken along two mutually perpendicular diameters.

   In the case of a micrometer microscope, care should be taken to avoid error due to back-lash while taking observations for the measurement of diameter.
5. Since Poiseuille's formula is true for streamline motion only, the pressure difference between the ends of the capillary tube should be small so that the liquid may leave the tube in a trickle.
6. For measurement of Q, a graduated cylinder of small least count should be used and sufficient quantity of water should be collected in it.
7. While measuring the rate of flow of the liquid through the tube, the pressure difference between its ends must be constant. Hence the level of water in the constant level water tank should be maintained constant.
8. If the level of water in the vertical glass tube, used to measure h, changes as the water trickles down from the end of the tube, a small clean glass rod may be placed in contact with the water at the end of the tube.
9. If the water runs back along the side of the capillary tube, a little grease or vaseline should be smeared at the end of the tube, in order to well clear of the bore.
10. (10)Since the viscosity of a liquid varies rapidly with temperature, the temperature of water should be observed with great care with a sensitive thermometer. In the neighborhood of 15°C, for water decreases by about 3 % for each degree rise of temperature.

Viva Voce

1. What is viscosity? Explain.
2. Explain the term viscous force.
3. Is viscosity same for all the fluids?
4. What is the effect of temperature on the viscosity of a liquid?
5. How does the microscope work in this set up?
6. What is streamline and turbulent flow of liquid?
7. Explain viscosity gradient.
8. What do you mean by velocity gradient?

Experiment No. 19

Co-Efficient of Viscosity of a Liquid

Object

To determine the co-efficient of viscosity of a liquid, say castor oil by Stokes' method.

Apparatus

Falling-sphere viscometer , steel ball bearings, a thermometer, a stop-watch, a watch-glass, a pen-knife and ascale.

Principle

Let us consider a very small sphere of radius a and density ρ, be allowed to fall freely from rest through a fluid of density σ and co-efficient of viscosity η. At any instant the sphere is acted upon by two forces, viz.,

A viscous drag $6\pi\eta aV$ directed vertically upwards, where v is the velocity of the sphere, and

The effective weight $= \frac{4}{3}\pi a^3(\rho-\sigma)\, g$ of the sphere directed vertically downwards.

The viscous drag increases with velocity V of the sphere and so long as it is less than the other force the velocity of the sphere increases. Finally when the viscous drag equals the effective weight of the sphere, there is no resultant force acting on the sphere and consequently it acquires a constant velocity, called the terminal velocity. Thus, if V is the terminal velocity of the sphere, we have

$$6\pi\eta aV - \frac{4}{3}\pi a^3(\rho-\sigma)g$$

when $$\eta = \frac{2}{3} \cdot \frac{a^2(\rho - \sigma)g}{V} \qquad (1)$$

If t is the time taken by a sphere to fall through a distance s. when its velocity is constant

$$V = s/t$$

$$\eta = \frac{2}{9} \cdot \frac{a^2 t(\rho - \sigma)g}{s} \qquad (2)$$

The above equation (2) can be used to calculate η.

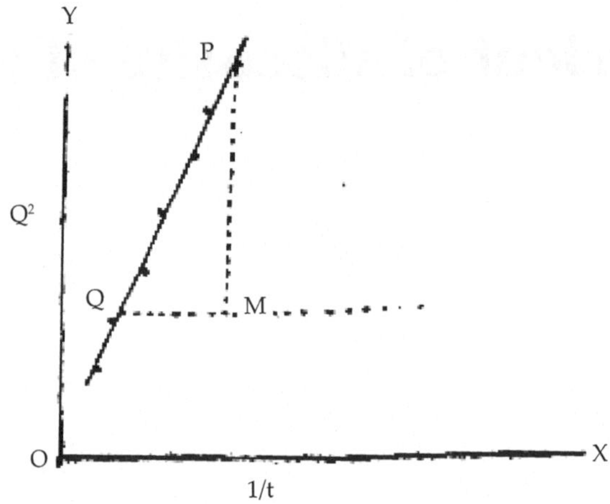

Fig. 1 plot of a^2 against l/t

If several spheres of different radii are allowed to fall through the liquid and their times of fall through the same height observed, a plot of a^2 against l/t should give a straight line as illustrated in fig. 1. The slope of this line will give an average value for a^2t which may be used in equation (2) to calculate η.

Description of Apparatus

The falling sphere viscometer arrangement is shown in Fig. (2). A is a glass tube containing the experimental liquid. It is about 50 cm long and 8 cm wide. This tube is closed at the top by a cork bored to carry centrally a narrow glass tube B of about 0.5 cm in diameter. It is through this tube B that the spheres are let fall into the liquid. This arrangement ensures that the motion of the spheres through the liquid is along the axis of the wider tube. The tube A is suspended in a large glass jar C containing water- into which dip a stirrer S and a thermometer T while F is a funnel through which water can be poured into the jar.

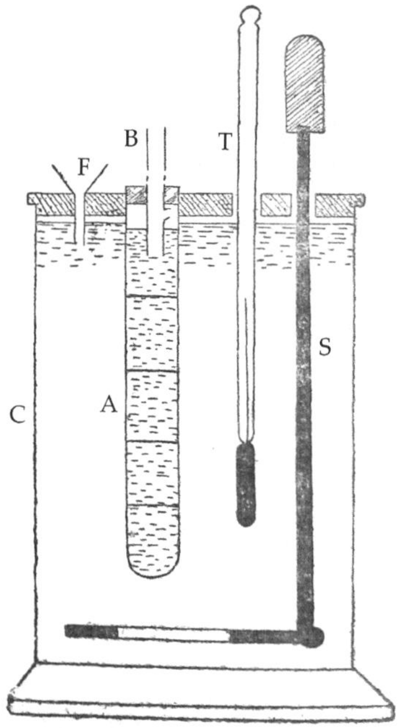

Fig. 2 Falling-sphere viscometer

Procedure

1. To begin with performing this experiment, take several steel ball-bearings of not more than 1/8″ in diameter and measure the diameter of each ball accurately with a screw gauge.
2. Then place the balls in the experimental oil M taken in a watch-glass and roll the mone by one until their surfaces are thoroughly wetted by the oil.
3. Set up of the viscometer is arranged as shown in Fig. 2 and four cotton threads are tied , preferably coloured, around the wide tube A dividing it into five portions, the three central ones being adjusted to be exactly 10 cm.
4. Then one ball - bearing is transferred on the point of a pen knife to the top of the narrow tube B. The ball, after passing through this tube, will fall centrally through the oil. When the ball reaches the first thread, start the stop-watch and note down the times taken by the ball to travel between successive threads, and thus find the mean value of the time of transit between any two consecutive threads. Then calculate the value of a2t for the ball.
5. The above experiment is repeated with other ball-bearings and as earlier, the mean time of fall of each ball between any two consecutive threads is found out. Calculate the value of a^2t for each ball separately and then take the mean. Use this mean value of a^2t in equation (2) to calculate the value of co-efficient of viscosity of the oil. The density of the oil, if not known, can be determined with the help of a hydrometer.

6. As a next step, a graph is plotted taking 1/t along the X—axis and a2 along the Y—axis. The graph will come out to be a straight line as shown in fig.1. The slope of this line PM/QM gives the average value of a^2t. Substitute this value in equation (2) and again calculate the value of V. Finally measure the temperature of water in the jar with correctness of up to 1/10ºC.

Observations and Calculations

[A] (i) Density of the oil = ...gm./c.c.

(ii) Density of the ball bearing = ...gm./c.c.

(iii) Temperature of the water-bath = ... ºC

[B] Measurement of diameter of ball-bearings

Least count of screw gauge = ... cm.

Zero error = ... cm.

TABLE 1

Ball	Diameter in any direction x cm	Diameter in perp. direction y cm	Mean $\frac{x+y}{2}$ cm	Radius a cm.	a^2 cm^2
A					
B					
C					
D					
....					
....					
G					
H					

[C] Measurement of time of fall

Least count of stopwatch = ... sec.

TABLE 2

Ball	Distance between consecutive threads cm.	Time of transit sec.	Mean distance s cm.	mean time of fall t sec.	1/t sec^{-1}	a^2t cm^2sec	Mean cm^2 sec
A							
B							
.....							
.....							
.....							
G							
H							

$$\eta = \frac{2}{9} \cdot \frac{a^2 t(\rho - \sigma) g}{s} = \ldots\ldots\ldots\ldots\ldots\ldots = \ldots\ldots\ldots\ldots\ldots \text{ poise.}$$

Average value of a^2t obtained from graph (Fig. 1)

$$\frac{PM}{QM} = \ldots\ldots\ldots\ldots\ldots\ldots = \ldots\ldots\ldots\ldots\ldots\ldots\ldots \text{ cm}^{2}\text{·sec.}$$

$$\eta = \frac{2}{9} \cdot \frac{a^2 t(\rho - \sigma) g}{s} = \ldots\ldots\ldots\ldots\ldots\ldots = \ldots\ldots\ldots\ldots\ldots \text{ poise.}$$

Result

The value of coefficient of viscosity of the liquid (..................) at .... ºC as obtained experimentally

(i) by calculation = ... dynes per cm^2 per unit velocity gradient.

(ii) by the use of graph = ... dynes per cm^2 per unit velocity gradient.

Sources of Error and Precautions

1. The ball-bearings should be of small size in order that their time of transit between any two consecutive threads may be sufficiently large to admit of accurate measurement.
2. In order that the wall-effect on the value of terminal velocity may be negligible, the radius 01 the tube containing the experimental liquid should be large as compared with the radius of the falling sphere.
3. The ball-bearings should be thoroughly wetted in the experimental liquid before they are allowed to fall through it and the wet bails should be picked up only by a pen-knife.
4. The balls should fall centrally into the experimental liquid.
5. As the viscosity of oil changes very rapidly with temperature the fall-experiments with various balls should be carried out one after the other as quickly as possible.
6. The radius of the ball occurs raised to the second power in the expression for η and is a small quantity. Hence the diameter of balls should be measured very accurately and in two mutually perpendicular directions.
7. The temperature of the liquid should be measured with a sensitive thermometer reading up to 1/10 °C.
8. The graph between a^2 and 1/t should be plotted on alarge graph paper and the scales be so chosen as to represent fully the observed accuracies in the quantities to be plotted.
9. The graph between a^2 and 1/t should be a straight line and the various points should be evenly distributed about it and should lie as close to it as possible.

Viva Voce

1. What is viscosity? Explain.
2. Explain the term viscous force.
3. Is viscosity same for all the fluids?
4. What is the effect of temperature on the viscosity of a liquid?
5. How does the microscope work in this set up?
6. What is streamline & turbulent flow of liquid? Also explain the term 'velocity Gradient'.
7. Explain viscosity gradient.
8. What is coefficient of viscosity?
9. What is the difference between capillary method & Stokes' method?
10. What is the least count of thermometer used in this experiment?
11. Give any application of this experiment.

Experiment No. 20

Magnifying Power of a Telescope

Object

To determine the magnifying power of a telescope.

Apparatus

Telescope, scale with markings and measuring tape.

Description

Magnifying power of a telescope is shown in Fig.1. The image formation through an astronomical telescope. The objective O forms the image of the object AB, at PQ, at a distance V from the objective. This image PQ serves as object for the eye-piece and forms a virtual magnified image at A'B'. Let us consider the distance of PQ from E be u and of A'B' from E be v. The final image may be formed either at the least distance of distinct vision or at infinity.

The magnifying power of a telescope is defined as the ratio of the angle subtended at the eye by the final image and the angle subtended at the eye by the object when viewed at its actual distance.

Let the distance of the object AB be U from the objective.

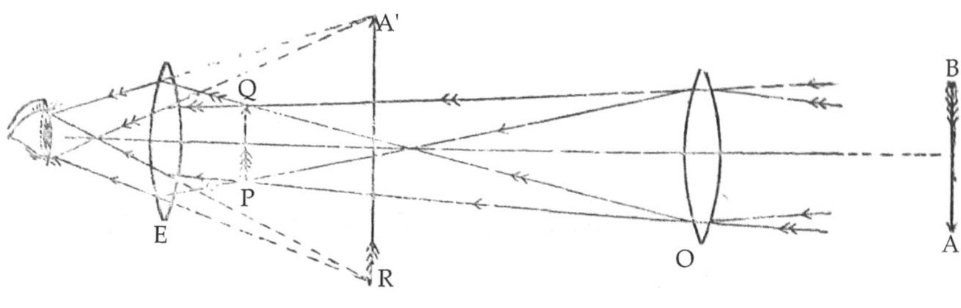

Fig. 1 Image formation through telescope

Hence angle subtended by the object=AB/U and the angle subtended by the image= A'B'/v

Therefore, magnifying power $$m = \frac{A'B'}{AB} \cdot \frac{U}{v}$$

But, considering the magnification produced by the objective,

PQ/AB=V/U

and, considering the magnification produced by the eye-piece,

A'B'/PQ=v/u

Hence $$A'B'/AB = \frac{V}{U} \cdot \frac{v}{u}$$

Substituting the value of A'B'/AB above, we have

m=V/u

(i) When the image is viewed at farthest point from the eye, the image PQ must be in the focal plane of the eye-piece, i.e., u=f.

m=V/f

If the object lies at infinity, V=F, m=F/f.

(ii) When the image is viewed at nearest point from the eye at the least distance of distinct vision, v=D, then

1/u = (D+f) / Df

and $$m = \frac{V}{f}\left(1 + \frac{f}{D}\right) \tag{1}$$

Principle

If a scale with markings quite wide apart, is viewed through the telescope, such that its image is formed at the same place, then, n divisions of the scale as seen through the telescope may coincide with n divisions of the scale viewed directly. Then the magnifying power is evidently given by

$$m = n/p$$

Procedure

1. Place a scale with markings at a distance of 20 ft. from the telescope. Look through the telescope with one eye, and directly at the scale with the other. This method presents difficulty in observation as the two eyes are being used independently. To overcome this difficulty, the eye-piece is so adjusted that the final image through the telescope is formed at the same distance as the scale.
2. When the adjustment is correct and both eyes are used, a slight movement of the head reveals that there is no relative shift between the two images.

3. Now the attention is focused on one division of the scale as seen through the telescope and observe the number of division is observed directly which corresponds with this.
4. The experiment is repeated by fixing the attention on 2, 3 .... divisions of the scale as seen through the telescope and finding the number of divisions observed directly which correspond to these respectively. Calculate the magnifying power in each case and find the mean value.
5. To check the results, measure the distance of the cross-wire from the objective and from the eye-piece. The ratio of the distances gives V/u, the magnifying power of the telescope.
6. Now the distance of the scale is altered from the telescope and the above experiment is repeated and it is observed that the magnifying power varies with the distance of the object.

Observation and calculations

TABLE 1

Sl. no.	Distance of scale from objective (cm)	No. of scale divisions		Magnifying power n/p	Mean value

Result

The magnifying power of the telescope for .... distance is ....

Sources of Error and Precautions

1. The divisions on the scale used in the experiment should be wide apart otherwise they may not be visible directly with the naked eye.
2. There should be no parallax between the scale and its image as seen through the telescope.
3. A short-sighted person habitually using the spectacles must perform the experiment with spectacles on the eyes.

Viva Voce

1. What do you mean by magnifying power of a telescope?
2. Explain how the image is formed in a telescope?
3. What is the least distance of distinct vision?
4. What is real and virtual image?
5. What is resolving power?

Experiment No. 21

Resolving Power of a Telescope

Object

To verify the expression for the resolving power of a telescope.

Apparatus

A telescope, a powerful source of monochromatic light, variable rectangular aperture, a sheet of plate glass coated with a tin foil and containing two fine parallel lines distant 1 to 2 mm and a microscope.

Resolving power of a Telescope

Let O be a monochromatic point source of light emitting light of wavelength λ. Let G be the objective of a telescope (Fig. 1). In practice the telescope objective alone limits the beam of light entering it but for the clarity of discussion, suppose a rectangular aperture AA kept in front of it. The incident wave-front MN is brought to a focus at P, the objective, and is seen magnified with the help of the eye-piece E. The image P is not a point image but its intensity curve is of the form p. Consider another point object O' ying to the right of O. Its image will also be formed in the focal plane but displaced to the left of P. However, as O′ approaches O, its image P′ in the focal plane approaches P and for one position of O', P' will be so near to P that they will be just resolved. This condition is obtained when the intensity patterns p′ and p are very close that the maximum of one falls over the first minimum of the other. It is this criterion which establishes the limit of resolution; if O' approaches O any further, the images overlap and cannot be distinguished as two.

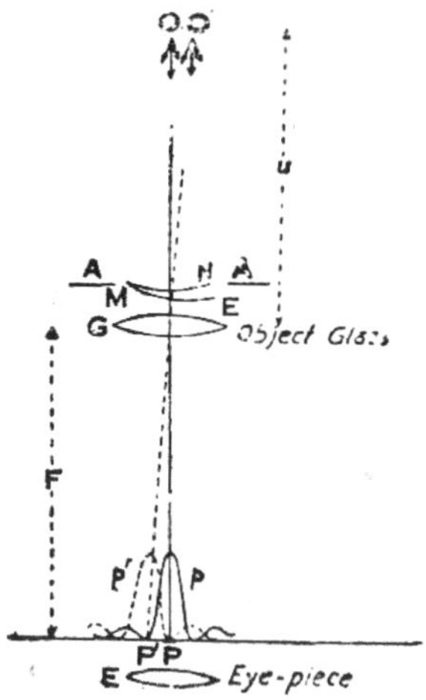

Fig. 1 Image formation through telescope

As the aperture AA is rectangular, the first minimum of the pattern p occurs when the distance NE=λ. Hence for the maximum of the pattern p' to fall on the first minimum of p, ME, the diffracted wave-front for O must be the actual wave-front from the object O′. The angle between the two wave-fronts is $\frac{NE}{MN}=\frac{\lambda}{d}$.

Where d is the width of the aperture. Since the rays are normal to the wave-front, the angle which the incident rays from O and O' subtend :At the objective is also $\frac{\lambda}{d}$. Incidentally this is also the angle subtended by image at the objective.If F be the focal length of the objective $\frac{OO'}{u}=\frac{\lambda}{d}=\frac{PP}{F}$ (1)

The relation above reveals that the theoretical least angle of resolution is the wavelength of light divided by the width of the aperture admitting; light into the telescope. The larger the value of d the smaller is the value of this angle and hence greater the resolving power of the telescope.

If the aperture is circular, accurate theory reveals that the first minimum surrounding the central disc occurs along the direction for which NE = 1.22A.

therefore, $$\frac{OO'}{u}=1.22\frac{\lambda}{d}$$

Supposing a telescope resolves two object points, it is then also necessary to use an eye-piece of sufficient magnification to see the images separately. The angle of resolution of the normal human eye is $1\frac{1}{2}$. Therefore the eye-piece must be of sufficient power to separate the beams of light forming the two images so that these leave the eye-piece at an angle of at least $1\frac{1}{2}$.

Theory

For a telescope fitted with a rectangular aperture the limiting angle of resolution θ, is given by

$$\theta = \frac{OO'}{u} = \frac{\lambda}{d}$$

where OO' =the distance between the two objects,

u =the distance of the objects from the objective of the telescope.

λ =wavelength of light coming from the objects,

d = width of the aperture, for just resolution of the two objects.

Procedure

1. A suitable object for this experiment is one that may be conveniently made by coating a plane glass sheet with a tin foil and cutting two parallel lines 1 to 2 mm apart. Mount this on a stand so that its plane is vertical and so also the two parallel lines.
2. Place a powerful source of sodium light behind this and focus the light on the slits. This plate then provides us with two bright linear sources of monochromatic light close together.
3. At a distance from this, adjust the telescope on a stand such that its axis lies horizontally and meets the plate providing the linear sources normally. By a horizontal movement of the telescope at right angles to its axis adjust that the images of the two slits lie symmetrically with respect to the intersection of the cross-wires in the field of view.
4. Place an adjustable slit (a rectangular aperture) as close to the objective of the telescope as possible. Then, diminish small steps the width of this aperture and thus adjust its minimum width for which the two images just to be resolved condition. Measure this minimum width of the slit with a micrometer microscope.
5. Also measure the distance between the plate providing the two linear sources and the object glass of the telescope.
6. Repeat the experiment for different distances of the plate providing the sources from the objective of the telescope. Finally measure the distance between the two linear sources with the help of a micrometre microscope.
7. For each distance, calculate the value of $\frac{OO'}{u}$ and $\frac{\lambda}{d}$ and compare the two results.

Observations and Calculations

TABLE 1

Sl. No.	Minimum width of slit for resolution d	Wavelength of Length λ	Theoretical resolving power λ/d	Distance between two linear sources OO'	Distance between target and objective u	Practical resolving power OO'/u	$\frac{OO'/\lambda}{u/d}$
1							
2							
3							
4							
5							

Result

The expression for the resolving power of the telescope has been verified.

Sources of Error and Precautions

1. The plane of the target consisting of two bright linear sources should be vertical.
2. The axis of the telescope should be horizontal.
3. The images of the two slits should lie symmetrically with respect to the intersection of the cross-wires.
4. Due care must be taken while measuring the width of the aperture and the distance between the two linear sources with a microscope. Avoid back lash error during the course of measurements made with it.
5. The minimum width of the aperture for resolution of the two objects should also be found by opening the slit.

Viva Voce

1. What is a monochromatic light?
2. What monochromatic light is being used in your experiment?
3. What is resolving power of a telescope?
4. Differentiate between resolving power with magnifying power.
5. On what factors does the resolving power of a telescope depend?
6. What is the magnifying power of a telescope?

Experiment No. 22

Searle's Method

Object

To determine the value of Y, n and σ for the material of a wire (..........) by Searle's method.

Apparatus.

Two equal inertia bars made of brass, about 30 cm., long and of square cross section of about 2 cm,, side with the necessary screws and clamps, a wire of about 30 cm., length and about 1 mm., diameter of the material to be tested, an accurate stop-watch, a screw gauge, a vernier callipers, thread, meter scale, a weight box and a balance.

Principle

Referring to fig. 1, two equal inertia bars AB and and CD of square section are joined at their centres by a fairly short and moderately thin wire GG′ of the material whose elastic constants are to be determined, and the system is suspended by two parallel torsion-less threads so that in the equilibrium position the bars may be parallel to each other with the plane ABDC horizontal. If the two bars be turned through equal angles in opposite directions and be then set free, the bars will execute flexural vibrations in a horizontal plane with the same period about their supporting threads.

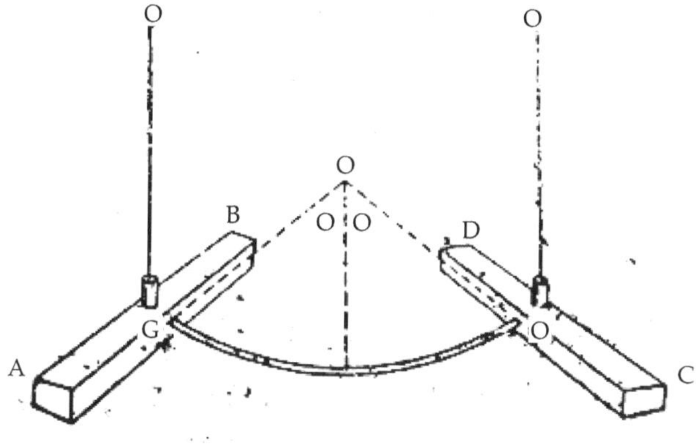

Fig. 1 *Searle's apparatus*

When the amplitude of vibration is small, the wire is only slightly bent and the distance GG' between the ends of the wire measured along the straight line will never differ perceptibly from the length of the wire so that the distance between the lower ends of the supporting threads remains practically constant and hence the threads remain vertical during the oscillations of the bars, and there is thus no horizontal component of tensions in the threads acting on the wire.

The mass of the wire is negligible as compared to that of the bars so that the motion of G and G' at right angles tb GG' may be neglected. Further since the horizontal displacements of G and G' are very small compared with the lengths of the supporting threads, the vertical motion of G and G' is also negligible. The centers of gravity of the bars, therefore, remain at rest, and hence the action of the wire on either bar and vice versa is simply a couple which, by symmetry, must have a vertical axis. The moment of this couple called the "bending moment" is the same at every point of the wire and thus the neutral filament of the wire is bent into a circular arc.

If ϱ is the radius of the arc, Y the Young's modulus for the material of the wire and I the geometrical moment of inertia of the area of cross-section of the wire about an axis through the centroid of the area and perpendicular to the plane of bending, the bending moment is from equation given by $G=YI/\varrho$. If l is the length of the wire and θ the angle turned through by either bar, $\varrho=1/2\ \theta$ and $G=\frac{2YI\theta}{l}$; and if $\frac{d^2\theta}{dt^2}$ is the angular acceleration of each bar towards its equilibrium position and K the moment of inertia of the bar about a vertical axis-through its C G., the torque due to inertial reaction is $K\frac{d^2\theta}{dt^2}$. Hence equating the sum of these two torques to zero, we get, from Newton's third law, the equation,

$$\frac{d^2\theta}{Kdt^2}+\frac{2YI\theta}{l}=0$$

or $$\frac{d^2\theta}{dt^2}+\frac{2YI\theta}{Kl}=0$$

The motion of the bars is, therefore simple harmonic and hence the period of flexural vibrations is given by,

$$T_1=2\pi\sqrt{\frac{Kl}{2YI}}$$

whence $$Y=\frac{2\pi^2 Kl}{T_1^2 I}$$

If the radius of the wire be r, $I=\frac{1}{4}\pi r^4$ and hence

$$Y=\frac{2\pi^2 Kl}{T_1^2 I} \qquad (1)$$

If the length and breadth (horizontal) of the bar be a and b respectively,

$$K=M\left(\frac{a^2+b^2}{12}\right) \qquad (2)$$

where M=mass of the bar.

Now the suspensions of the bars are removed and one of the bars is fixed horizontally on a suitable support (Fig. 2), while the other is suspended from the now vertical wire. If the wire is twisted through an angle and the bar allowed to execute torsional oscillations, the period of oscillation is given by,

$$T_2=2\pi\sqrt{K/c}$$

Where c is the restoring couple per unit radian twist due to torsional reaction of the wire and is equal to $\pi n r^4/2l$, where n is the modulus of rigidity for the material of the wire. Thus,

$$c=\frac{\pi nr4}{2l}=\frac{4\pi 2K}{T_2^2}$$

whence $$n=\frac{8\pi KL}{T_2^2 r^4} \qquad (3)$$

Dividing eqn. (1) by (3), we get

$$\frac{Y}{n}=\frac{T_2^2}{Y_1^2} \qquad (4)$$

NowY=2n(1+ σ), where σ is the Poisson's ratio. Hence

$$\sigma=\frac{Y}{2n}-I$$

or
$$\sigma=\frac{T_2^2}{2T_1^2}-I \tag{5}$$

Procedure

1. Take two cotton threads of about 60 cm, length and by means of them suspend the two inertia bars from a rigid support so that the axis of the bars and the wire, when the system is at rest, lie in the same horizontal plane as illustrated in Fig.1.
2. Then set up a pointer close to the end of one of the bars and make a mark on the bar in line with the pointer when the bar is at rest.
3. Now draw the two ends B and D of the bars slightly together by a loop of thin cotton thread slipped over them and bring the system thus constrained carefully to rest. Then burn the thread, the bars will be released and commence to oscillate regularly.
4. The amplitude of vibration should be small, about 30. Note down twice with an accurate stop-watch the time for a large number of oscillations, say 20 and the value of the period T1.Take at least two sets of observations for T1 with different number of oscillations and from them calculate the mean value of T1.

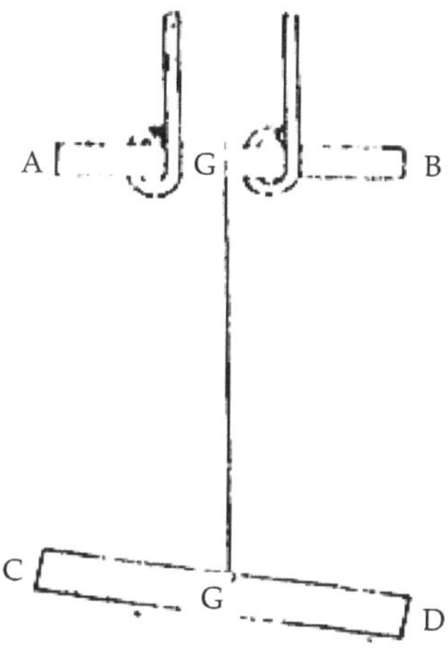

Fig. 2 Experimental arrangement for torsional oscillations

5. Now, remove the cotton threads supporting the bars and clamp one of the bars to a suitable support as shown in Fig. 2 so that the wire is vertical. Twist the wire by rotating the bar at its end in a horizontal plane and then release it. The bar will execute torsional oscillations.

Note down twice the time taken by the bar to complete again a large number of oscillations, say 20, and from it calculate the value of T2. Take at least two sets of observations for T_2 with different number of oscillations and from them find out its mean value.

6. Now measure with a screw gauge the diameter of the wire and thus find out the mean radius r. Next measure the length of the wire and the lengths of the inertia bars by a scale. Now measure the breadth of each bar with vernier callipers.
7. Finally weigh the bars separately in a balance and determine the masses. For very accurate work mass of each bar is found before the hole-is bored in it and before the hook is fixed to it, and the value of this mass is stamped on the bar.
8. Calculate the moments of inertia of each bar from equation (2) and thus find the mean value of K. Having determined K, l, r and the periods T1 and T_2, calculate the values of Y, n and σ from equations (1), (3) and (5).

Observations

(A) Determination of T_1 and T_2

Least count of stopwatch = … sec.

TABLE 1

Sl. no.	No. of oscillations	Determination of T_1				Determination of T_2			
		Time taken		T_1 Sec.	Mean	Time taken		T_2 Sec.	Mean
		Min.	Sec.			Min.	Sec.		
1									
2									
3									
4									

(B) (i) Length of the wire = ….. cm.

(ii) Measurement of diameter if the wire

Least count of screw gauge = ….. cm.

Zero Error = ….. cm.

TABLE 2

Sl. No.	Reading along any diameter a cm	Reading along perpendicular diameter b cm	$\frac{a+b}{2}$ cm	Mean

(C) (i) Mass of the inertia bar = ... cm.

(ii) Length of the inertia bar = ... cm.

(iii) Breadth of the inertia bar = ... cm.

Calculations

Mean radius of the wire = ... cm.

$$K = M\left(\frac{a^2 + b^2}{12}\right) = \text{.........=.........} \text{ gm. cm}^2.$$

$$Y = \frac{8\pi KL}{T_1^2 r^4} = \text{.........=.........} \text{ dynes/cm}^2.$$

and $$\sigma = \frac{T_2^2}{2T_1^2} - I = \text{.........=.........}$$

Result

The values of elastic constants for the material of the wire (.................) are

Y = ... dynes/cm^2.

n = ... dynes/cm^2.

σ = ...

Sources of Error and Precautions

1. While taking observations for period T1 of flexural vibrations, the amplitude of oscillation should be small. In that case threads will remain vertical and there will then be no horizontal component of the tension in the threads acting on the wire. Further the wire will also be not strained beyond elastic limit.
2. All undesirable motions (up and down) should be completely checked.
3. The periods occur raised to the second power in the expression for Y, n and σ. Hence they must be determined very accurately by timing a large number of oscillations with an accurate stop-watch up to the smallest value measurable with it.
4. While determining period T2 for torsional oscillations, the amplitude of oscillation may be kept quite large, but the wire should not be twisted beyond elastic limit.
5. Since the radius of wire occurs in the fourth power in the formula for calculation of elastic constants and is a small quantity, readings for diameter should be taken with a screw gauge at a number of points equally spaced along the wire, two perpendicular diameters being measured at each place taking care not to compress the wire while making measurements.

Viva Voce

1. Name and explain elastic constants.
2. Explain the term bending moment.
3. What do you mean by center of gravity of a body?
4. What is inertia of a body? Also explain moment of inertia.
5. What are torsional oscillations?

Experiment No. 23

Surface Tension by Jaeger's Method

Object

To determine the surface tension of a given liquid (   )by Jaeger's method.

Apparatus

Jaeger's apparatus, a glass tube of about 5 mm, diameter, a microscope, a scale and a thermometer.

Description of apparatus

As illustrated in Fig. 1, the Jaeger's apparatus consists of a thin glass tube drawn out at its lower end to a fine capillary and fixed vertically in the experimental liquid. The glass tube is connected to a manometer M and also to a bottle fitted with a dropping funnel (or burette) containing water.

Principle

When the capillary tube is dipped in the liquid, the liquid will Hit in it and the shape of the meniscus will be approximately spherical. If air is forced into the glass tube by dropping water from the funnel (or burette) into the bottle, the surface of the liquid in the capillary tube gets pressed downwards and as the pressure of air inside the tube increases the liquid surface goes on sinking lower and lower until finally a hemispherical bubble of radius r which is equal to that of the orifice of the capillary tube protrudes into the liquid below. The pressure inside the bubble being greater than that outside by 2T/r, where T is the surface tension of the liquid.

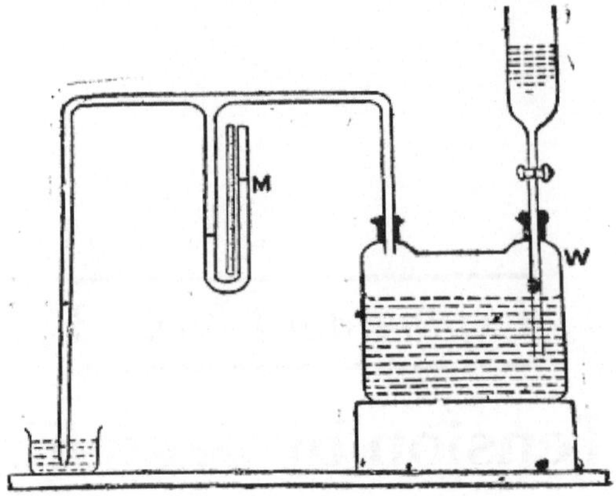

Fig. 1Experimental Setup for Jaeger's method

At this stage the pressure in the manometer reaches to a maximum and the bubble becomes unstable, for a slight increase in the radius decreases-the internal pressure necessary to produce equilibrium, and as the external pressure is constant, the bubble breaks away. Just before the bubble breaks away, the pressure inside it is Hgd + P, where H is the maximum reading of the manometer, d the density of the liquid in the manometer and P the atmospheric pressure. The pressure outside the bubble is hgϱ + P, where h is the depth of the tip of the capillary tube below the surface of the liquid and p the density of the liquid. Thus the excess pressure inside the bubble is,

$$(Hgd+P)-(hg\rho+P)=\frac{2T}{r}$$

Whence $$T=\frac{rg}{2}(HD-h\rho) \quad (1)$$

from which surface tension T of the liquid may be calculated.

Procedure

1. To begin with the experiment, take a thin glass tube, and at its one end draw out a capillary of about 0.2 to 0.5 mm diameter. Then clamp the tube in a vertical position, with the orifice of the capillary tube at about 3 or 4 cm below the surface of the experimental liquid contained in a beaker.
2. The lower end of the capillary is cut off square, i.e. perpendicular to the axis of the tube, without irregular inner edges when viewed under a microscope.
3. Now a scratch on the tube is made and the position of the tube is adjusted such that the scratch coincides with the liquid surface. This ensures that the bubbles are formed at the same depth h below the surface of the liquid.
4. Then the rest of the apparatus is arranged as shown in Fig. 1. If the apparatus is air-tight, the manometer will at once show a difference of pressure between the inside and the outside of the apparatus.

5. Now the air is blown into the top of the tube from the bottle by allowing water to drop into it from the dropping funnel and adjust the rate of flow of water into the bottle so that a bubble is formed about every ten seconds at the orifice of the capillary tube. The pressure indicated by the manometer rises until the bubble becomes unstable and then suddenly falls when the bubble breaks away.
6. The maximum difference H in heights of liquid in the two limbs of the manometer is noted down. The above experiment is repeated several times and the mean value of H is found out.
7. Then measured the depth h of the orifice of the capillary tube from the surface of the liquid. Now alter the value of h and take at least three sets of observations for H with different values of h. Repeat the experiment with two more capillary tubes of slightly different cross sections.
8. As a next step clamp the capillary tube in a horizontal position and focus a vernier or micrometer microscope on its orifice. Then coinciding the cross-wire with the inner edge of the tube at one side of the centre and then with its inner edge on the other side of its centre, determines the diameter of the orifice.
9. Now the tube is roated through a right angle and again its diameter is determined. In this way take several observations for two mutually perpendicular diameters which will be practically the same if the bore of the tube is circular, and their mean value will denote the value of 2r. Then calculate the value of surface tension of the experimental liquid for each set of observation separately and find out the mean value of surface tension. Finally the temperature of the experimental liquid is noted down.

Observations

(A) Measurement of h and H.

TABLE 1

Set	Depth of orifice h cm.	Manometer Readings (cm)			Mean H cm.
		Upper level of liquid	Lower level of liquid	H	
1					
2					
3					

(B) Temperature of experimental liquid = ... ^{0}C

(C) (i) Density of experimental liquid = ... gm/cc.

(ii) Density of manometric liquid = ... gm/cc.

(D) Measurement of diameter of orifice.

Least count of the microscope = ... cm.

TABLE 2

Sl No.	Observation for position of cross-wire at the section of the orifice of capillary tube						$\frac{a+b}{2}$ cm.	Mean Diameter Cm
	Reading along any diameter			Reading along prep. diameter				
	At one end of section cm.	At the other end of section cm.	Diameter a cm.	At one end of section cm.	At the other end of section cm.	Diameter b cm.		
1								
2								
3								

Calculations

Mean radius of the orifice of the capillary tube = ... cm.

Set 1 $T = \frac{rg}{2}[Hd - h\rho]$ = ... = ... dynes/cm.

Therefore, Mean value of surface tension = ... dynes/cm.

Result

The surface tension of the experimental liquid (.............) at ....ºC = ... dynes/cm.

Standard Value = ... dynes/cm.

Error = ... %

Sources of Error and Precautions

1. There should be no leakage in the apparatus. It is distinctly advantageous to have the apparatus in one piece avoiding the use of rubber joints.
2. The manometer should contain xylol[(CH3)2C6H4] so that the height H may be large ; the density of xylol is less than that of water.
3. To damp the oscillations of the liquid in the manometer, its open end may be drawn out to a capillary tube.
4. Diameter of the capillary tube must be measured only at the orifice dipping in the liquid.
5. The orifice of the capillary tube should be circular and very small, about 0.3 mm., in diameter, so that the maximum pressure in a bubble may occur when it is hemispherical.

6. While measuring the diameter of the orifice of the capillary tube, several observations of mutually perpendicular diameters should be taken. This reduces the error due to ellipticity of the cross-section to minimum.
7. In the case of a micrometer microscope error due to back-lash should be avoided while taking: observations for diameter.
8. The capillary tube should be clean. All traces of grease must be carefully removed as they are very fatal to surface tension experiments.
9. The bubbles should be formed singly and slowly say at the rate of one in about ten seconds, otherwise the maximum pressure registered by the manometer will not be independent of the rate of bubble formation. The first condition is satisfied by keeping the air space in the Woulf's bottle small so that when one bubble breaks away from the end of the capillary tube, the pressure inside the apparatus is reduced appreciably below the maximum required to blow the bubble. The second condition is fulfilled by adjusting the flow of liquid from the dropping runnel into the Woulf's bottle.
10. As the surface tension of a liquid depends upon its temperature, the temperature of the experimental liquid should be recorded.

Viva Voce

1. What is surface tension?
2. Is surface tension same for all liquids? If not why?
3. How surface tension varies with temperature?
4. What is manometer?
5. Write down the unit of surface tension.
6. What are the cohersive and adhesive forces?

Experiment No. 24

Surface Tension of Water

To determine the surface tension of water by using capillary tube.

Apparatus

Glass tubes to draw capillary tubes, a beaker, a clean glass plate, a micrometer eye-piece microscope and a thermometer.

Theory

Let a capillary tube dip in a liquid of surface tensior T, the axis of the tube being vertical, and let ABC be the surface of the liquid in the tube. If the cross-section of the tube at the meniscus ABC inside the tube be a circle of uniform internal diameter 2r, the liquid meniscus touches the inner surface of the tube round the circumference of a circle of radius r, that is round a length $2\pi r$; and the total pull exerted by the surface film or is the glass is equal to $2\pi r$ T.

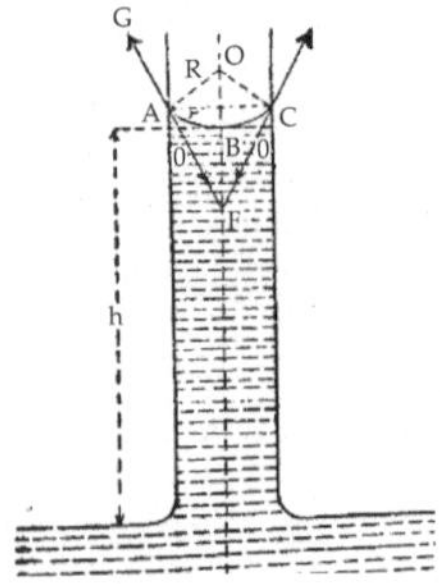

Fig. 1 Capillary Rise Method

If θ be the angle of contact between the liquid and the glass, the tangent plane to the liquid surface at any point A will make an angle θ with the vertical and hence the pull on the glass at any point A will be along AF. Since action and reaction are equal and opposite, the glas exerts an equal and opposite pull on the liquid film along AG. The vertical component of this reactional force on unit length of the line of contact is T cos θ and the horizontal component is T sin θ outward. Considering, the whole meniscus, the horizontal components cancel out while the vertical components are added up and the resultant force on the film along the entire line of contact at the meniscus acts upwards and is $2\pi r T\cos\theta$. This force supports the weight of the corresponding column of liquid raised above the level outside.

If h is the height of the capillary ascent, measured vertically from the bottom of the meniscus to the horizontal part of the free surface outside, the volume of the liquid column supported inside the tube blow the horizontal tangent plane to the meniscus is $\pi r^2 h$. If the bore of the tube is very fine, the meniscus is a hemisphere of radius r and hence the volume v of the liquid above the horizontal tangent plane to the meniscus is approximately equal to the difference between the volume of a cylinder (Fig. 2) of radius r and length rand the volume of a hemisphere of radius r, or

$$v = \pi r^2 . r - \frac{2}{3}\pi r^3 = \frac{1}{3}\pi r^3$$

and hence the total volume of the liquid column supported inside the tube is

$$\pi r^2 h + \frac{1}{3}\pi r^3$$

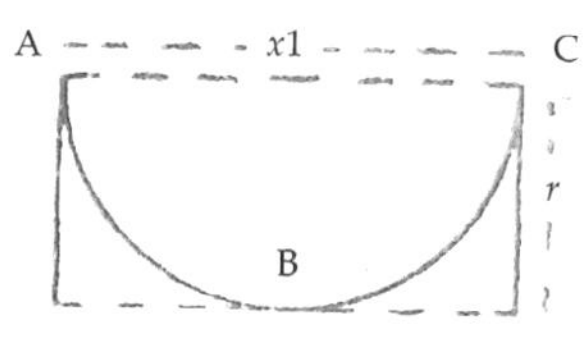

Fig. 2

If the density of the liquid be P and the acceleration due to gravity gy the weight of the column of liquid is $2\pi r^2(h+r/3)\rho g$ Equating this to the resultant upward force exerted on liquid film by the glass, we get

$$2\pi r T\cos\theta = 2\pi r^2(h+r/3)\,\rho g$$

whence $$T = \frac{r(h+r/3)\rho g}{2\cos\theta} \qquad (1)$$

Procedure

1. Take several pieces of soft glass tubing about 0.2 to 0.5 cm. in diameter and about 10 cm in length and clean them by rinsing first with caustic soda, then with nitric acid, and then with considerable quantities of tap water to remove all traces of nitric acid. Then dry the tubes by drawing or forcing a current of warm air through them.

2. Now heat one of the clean glass tubes near the middle in a Bunsen-or blowpipe flame rotating the tube slowly all the time to secure equal heating on all sides. When the glass-begins to soften apply a gentle-pressure along its length so that the walls of the tube thicken.
3. Then remove it from the flame and draw the two ends apart, at first slowly and then more quickly, until a capillary tube of about ¼mm to 1 mm in internal diameter and of about 30cm length is formed. Cut off from the central portion of this a piece of about 20 cm., length. In this way prepare at least half-a-dozen capillary tubes and from them select thr or four tubes of uniform bore.
4. As a next step, take a clean thin rectangular glass plate about 6" long and 3" wide and attach to its lower end a small pin parallel to its length. Lay on the plate one of the capillary tubes, with its length parallel to the length of the plate, and fix it in position by two rubber bands or wax.

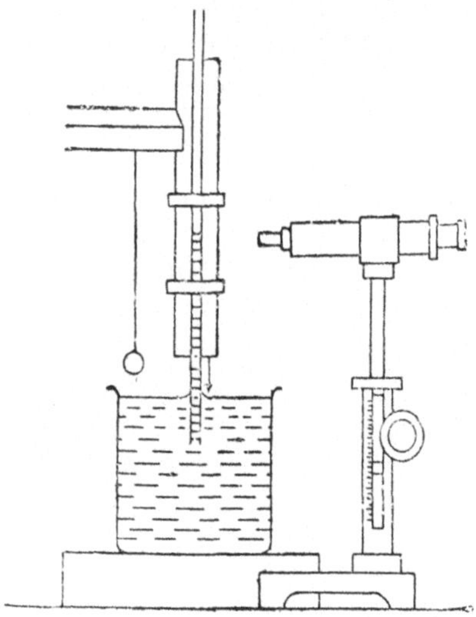

Fig. 3 Experimental Setup

Then hold the plate vertically in a clip as depicted in fig. 3, so that tip of the pin is just not in contact with the free surface of water contained in a clean glass beaker and the capill-larytube is about 3 inches under the surface of the liquid. The water will rise in the capillary tube. When the water has ceased rising, slowly raise the glass plate about two inches. If the water falls back readily as the tube is raised, the tube and the liquid are not contaminated. This also ensures that the bore of the tube in the neighbourhood of the meniscus is wetted. Then clamp the plate so that the capillary tube and the plate as judged by a plumb line are vertical.

Now to measure hfocus the cat hetometer microscope on the liquid meniscus inside the capillary tube and note down its reading. Then lower down the microscope and focus it on the tip of the pin, the microscope being set so that when images of the pin and its reflection are viewed through it, the cross-wire is exactly between them. Again note down the reading of the microscope. The difference between the two readings of cathetometer microscope gives the value of h. Next make a mark with ink on the outside of the tube at the position of the meniscus and then removing it from the plate make a scratch on the tube at the ink-mark with a hardened steel knife or a file and by applying tension along its axis break the tube into two pieces. The tube should not be broken by bending it otherwise the fracture will be uneven and not perpendicular to the axis.

Next fix one of the two broken parts vertically on a piece of plate glass with red wax, with the broken end uppermost. Then focus the micrometer eye-piece microscope on the end of the tube and set the cross wire tangential to the internal boundary line of the section and take the reading of the micrometer attached to the eye-piece. Now move the cross-wire at right angles to its length until it coincides with the internal boundary of the section on the other side of the centre and again take the reading of the micrometer. The displacement of the crosswire as known from the difference between the two readings of the micrometer gives the diameter of the tube. Repeat the observations several times and find out the mean. Now turn the tube through a right angle about its axis and again take several readings for the diameter and calculate its mean value. Similarly find out the two mutually perpendicular diameters of the broken end of the other part of the capillary tube. The mean of the values for the four diameters, two mutually perpendicular diameters for each broken part, if differing very little, should be taken as zr.

Having thus determined h and r, calculate the value of surface tension T from the equation (1). Repeat the experiment with a few more capillary tubes and calculate the mean value of surface tension. Lastly note down the temperature of water.

Observations

[A] Determination of b.

Least count of cathetometer microscope = ... cm.

TABLE 1

No. of capillary tube	Reading of meniscus (cm)	Reading of tip of pin (cm)	b (cm)
A			
B			
C			

[B] Determination of diameter: Capillary tube No. A.

Least count of cathetometer microscope = ... cm.

TABLE 2

<table>
<tr><th rowspan="3">Sl. No.</th><th colspan="12">Observation for position of cross-wire at the section of the tube</th></tr>
<tr><th colspan="6">Broken end of one piece</th><th colspan="6">Broken end of other piece</th></tr>
<tr><th colspan="3">Reading along any diameter</th><th colspan="3">Reading along perp. diameter</th><th colspan="3">Reading along any diameter</th><th colspan="3">Reading along perp. Diameter</th></tr>
<tr><th></th><th>At one end of section (cm)</th><th>At the other end of section (cm)</th><th>Diameter (cm)</th><th>At one end of section (cm)</th><th>At the other end of section (cm)</th><th>Diameter (cm)</th><th>At one end of section (cm)</th><th>At the other end of section (cm)</th><th>Diameter (cm)</th><th>At one end of section (cm)</th><th>At the other end of section (cm)</th><th>Diameter (cm)</th></tr>
<tr><td>1
2
3</td><td></td><td></td><td></td><td></td><td></td><td></td><td></td><td></td><td></td><td></td><td></td><td></td></tr>
<tr><td colspan="4">Mean diameter =.......cm</td><td colspan="3">Mean diameter =.......cm</td><td colspan="3">Mean diameter =.......cm</td><td colspan="3">Mean diameter =.......cm</td></tr>
</table>

[C] Temperature of water = ... ºC

Calculations

For capillary tube A.

Mean diameter of the tube = ... cm

Mean radius of the tube = ... cm

$$T = \frac{r(h + r/3)g}{2} = \ldots\ldots\ldots\ldots\ldots = \ldots\ldots\ldots\ldots \text{ dynes/cm.}$$

Make similar calculations for surface tension from observations with other capillary tubes.

Therefore, Mean value of surface tension of water= .... dynes/cm.

Result

The surface tension of water at °C = ... dynes / cm.

Standard value = ... dynes / cm.

Error = ... %

Sources of Error and Precautions

1. The rise of a liquid in a capillary tube depends upon the radius of the tube at the liquid meniscus. Hence the diameter of the tube should be determined at the meniscus only. Further the tube should be of uniform bore otherwise the measurement of diameter even slightly below or above the meniscus

may be a source of considerable error in the result. The uniformity of the bore may be tested by introducing a thread of mercury in the tube and measuring the length of the thread in different positions which, if the bore is uniform, will be the same.

2. While determining the diameter of the tube with a micrometer eye-piece microscope, error due to back-lash should be avoided by turning the screw always in the same direction before taking a reading. Two mutually perpendicular diameters should be measured at the broken end of each of the pieces of the capillary tube.
3. While observing the rise of liquid in a capillary tube, the tube must be vertical as judged by a plumb line.
4. The top of the capillary tube should be open and not blocked by wax, etc.
5. The capillary tube, the glass plate and the beaker should be clean and not contaminated with oil or grease. Tap water should always be used as distilled water is very-apt to have a minute trace of grease.
6. As the surface tension of a liquid varies with temperature, the temperature of water should always be recorded.

Viva Voce

1. What is surface tension?
2. Is surface tension same for all liquids? If not why?
3. How surface tension varies with temperature?
4. What is manometer?
5. Write down the unit of surface tension.

Experiment No. 25

Poisson's Ratio

Object

To determine the Poisson's ratio for rubber.

Apparatus

Rubber tube with metal sleeve and rubber stopper, burette, 250 gm, slotted weights, metre scale and a pointer.

Theory

Let us consider a rubber tube suspended vertically to be loaded at its lower end with a small weight. This stretches the rubber tube slightly with a consequent increase in its length and internal volume. Let V be the original volume of the tube, A its area of cross-section, L its length and D its diameter. Then, if for a small increase of volume dV, the corresponding increase in length is dL and the decrease in area is dA, then we have

$$V + dv = (A\text{-}dA)(L\text{+}dL)$$

$$= AL\text{+}A.dl\text{-}L.dA\text{-}dA.dL$$

Putting AL=V and neglecting dA. dL, a product of two small quantities, we have

$$dv = A.dL\text{-}L.dA \qquad (1)$$

Now $\qquad A = \pi(D/2)^2 = \pi D^2/4$

Differentiating $\qquad dA = \frac{\pi D}{2}.dD = \frac{2A}{D}dD$

Substituting this value of dA in eqn. (1), we get

$$dV = A.dL - \frac{2AL}{D}.dD$$

or

$$\frac{dV}{dL} = A - \frac{2AL}{D}.\frac{dD}{dL}$$

$$\frac{L}{D}.\frac{dD}{dL} = \frac{1}{2}\left[I - \frac{IdV}{AdL}\right]$$

or

$$\sigma = \frac{1}{2}\left[I - \frac{IdV}{AdL}\right] \qquad (2)$$

This equation can be employed to calculate σ, if other quantities are determined.

Procedure

1. A rubber cycle-tyre tube AB of about one metre in length suspended vertically as shown in Fig.1. Close it securely at its upper end with a rubber stopper and seccotine, and at its lower end with a metal sleeve carrying a hookH. Bore the stopper at the upper end of the tube to carry a burette C held vertically.
2. Set up a meter scale S vertically in front of the tube and attach a small horizontal pointer P to the metal sleeve B so as to be just no tin contact with the scale.
3. Water is poured countinously into the burette until the whole of the rubber tube and about 4/5 of the burette are filled with it. The apparatus is then allowed to settle down for about 5 minutes. Measure the diameter of the rubber tube at a number of points with a vernier callipers and find its mean value, and thus the area of cross-section of the tube is calculated.
4. As a next step the position of the water meniscus inside the burette and the position of the pointer P on the scale S is noted down. Place gently a load, of say 250 g, on the hanger at the lower end of the tube and wait for about 5 minutes.
5. Again the readings of the burette and the pointer P on the scale S are taken. The difference between the two readings of the burette gives the change in volume dV for the load on the hanger and the difference between the two readings of the pointer P on the scale gives the corresponding change in length.
6. Now, the load on the hanger is increased in equal steps of, say 250 g, till maximum permissible load within elastic limit is reached, taking down the readings of the burette and the pointer P after addition of each load when the apparatus has settled down.
7. Then the load on the hanger is decreased in the same stages and the above observations are repeated till the load on the hanger is reduced to Zero.

Take the mean of the two readings of the burette for the same load on the hanger obtained with increasing and decreasing load ; and then subtracting the mean reading for zero load on the hanger from the mean reading for any load, calculate the change in volume dV of the rubber tube for various loads on the hanger.

8. Similarly calculate the value of corresponding change in length dh for the various loads. Then the value of Poisson's ratio σ for rubber is calculatd from equation (2) for each set of observations separately and the mean value of σ is found out.

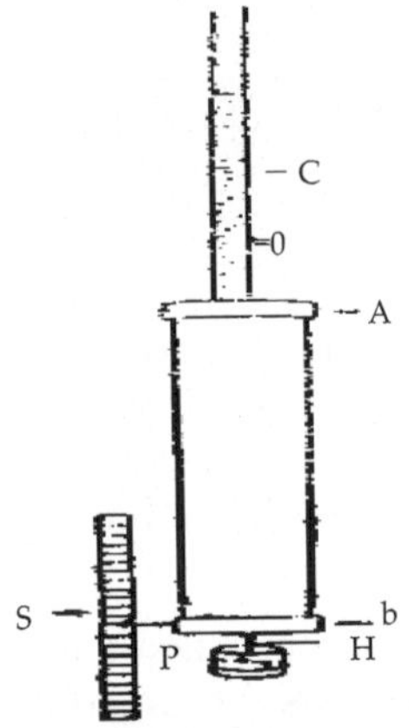

Fig. 1 Experimental Setup

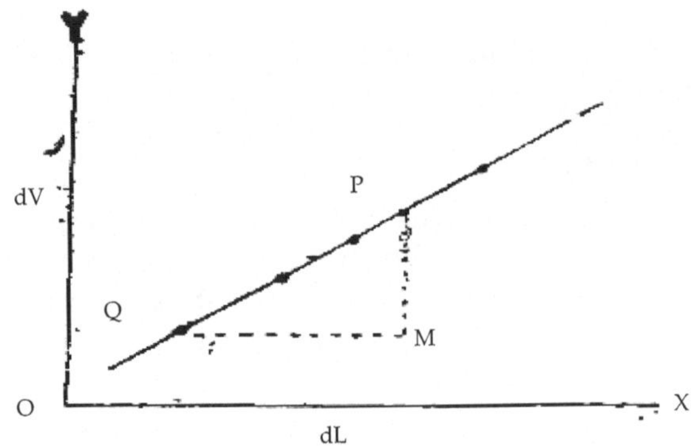

Fig. 2 Slope curve

9. Next step is to plot a graph taking dL along the X-axis and the corresponding values of dV along the Y-axis. This will come out to be a straight line PQ as shown in Fig. 2, and its slope PM/QM will give the average value of dV/dL. Use this in equ. (2) to calculate again the value of σ.

Observations

(A) Measurement of diameter of the rubber tube :

Vernier Constant = …… cm.

Zero Error (if any) = ……. cm.

TABLE 1

Sl. No.	Reading along any diameter a cm	Reading along perpendicular diameter β cm	$\frac{\alpha+\beta}{2}$ cm	Mean uncorrected diameter
1				
2				
3				
4				
5				

(B) Determination of change in volume dV and the correspnding change in length dL:

TABLE 2

Sl. No.	Load on hanger (gm)	Reading of burette (cc)			Change in volume dV (cc)	Reading of pointer P on scale S (cm)			Change in length dL (cm)	Poisson's rati σ
		Load increasing x	Load decreasing y	Mean $\frac{x+y}{2}$		Load increasing x′	Load decreasing y′	Mean $\frac{x'+y'}{2}$		
1	0			a				a′		
2	250			b	b-a			b′	b′-a′	
3	500			c	c-a			c′	c′-a′	
4	750			d	d-a			d′	d′-a′	
5	1000			e	e-a			e′	e′-a′	
6	1250			f	f-a			f′	f′-a′	

Mean: ....

Calculation

Mean corrected diameter of the rubber tube = ... cm.

Therefore, Area of cross section of the tube, $A = \pi D^2/4 =$ .....sq. cm.

I set : $\sigma = \frac{1}{2}\left[I - \frac{I}{A}.\frac{dV}{dL} \right]$

The mean value of Poisson's ratio for rubber = ...

From Graph PM = ... c.c. QM = ... cm.

Therefore, $\frac{dV}{dL} = \frac{PM}{QM}$ ..... sq. cm.

and $\sigma = \frac{1}{2}(-A.I.\frac{dV}{dL}) =$......................

Result

The value of Poisson's ratio for rubber as obtained experimentally,

(i) by calculations = ...

(ii) by graph = ...

Standard value of σ = ...

Therefore, Error = ...

Sources of Error and Precautions

1. The vertical scale setup in front of the tube should be arranged to give maximum extension allowable and the pointer should be just not touching the scale.
2. There should be no air bubble inside the rubber tube or the burette.
3. Diameter of the rubber tube should be measured at a number of points along its length and at each point two diameters at right angles to each other should be measured.
4. Weights should be placed or removed gently and in equal steps of, say 250 gm.
5. After each addition or removal of load wait for about 5 minutes before taking observations in order to allow the apparatus. to settle down to new conditions of stress and strain.
6. The load suspended at the lower end of the rubber tube should not exceed the maximum load permissible within elastic limit.

Viva Voce

1. What is elasticity? Explain the term elastic limit & Hooke's law.
2. What is stress and strain?
3. What is volumetric strain?
4. Differentiate between elasticity and plasticity.

Experiment No. 26

Sonometer

Object

To determine the frequency of Alternating Current (A.C.) mains by means of a sonometer.

Apparatus

Sonometer, half-kilogram weights, horse-shoe magnet, clamp stand, step-down transformer, metre scale and a screw gauge.

Theory

Let us consider a sonometer wire stretched under a constant load to be placed in a uniform magnetic field applied at right angles to the wire in the horizontal plane and let an alternating current of low voltage be is allowed to pass through the wire. On account of interaction between the magnetic field and the current in the wire, the wire will be deflected, the direction of deflection being given by Fleming's left-hand rule. Since the current passed is alternating, therefore for half the cycle the wire will move upwards and for the next half cycle it will move downwards. In other words, the wire will receive impulses alternately in opposite directions at the frequency of the current. As a result of this the wire will execute forced vibrations with a frequency f of the A.C.

Now if the distance between the two bridges of the sonometer is so adjusted that the natural frequency of vibration n of the wire equals to that of the current, there will be resonance and the wire will vibrate with a large amplitude. At this stage

$$f = n = \frac{1}{2l}\sqrt{\frac{T}{m}} \tag{1}$$

Where l is the length of the wire between the bridges, m its mass per unit length and T is the tension in the wire.

$$\mu = \pi r^2 \rho \tag{2}$$

where r is the radius of the wire and ρ the density of its material. Thus, if m is evaluated from equation (2) whereas the frequency of alternating current can be calculated from equation (1).

Procedure

1. A small horse-shoe magnet N is mounted, vertically at the centre of a sonometer wire such that the wire passes freely in between the poles of the magnet and the face of the magnet is normal to the length of the wire.
2. Now a suitable tension to the wire, say of 4 kilograms weight is applied. The voltage of the A.C, mains is reduced to about 6 or 8 volts by means of a step-down transformer and then this reduced voltage is applied between the two ends of the sonometer wire.
3. As the current passes through the wire, it will begin to vibrate. The vibrations of the wire can be observed by naked eye or if virations are too small they can be felt by touching the wire with a finger.
4. Now the distance between the two bridges is altered so that the amplitude of vibration of the wire increases and finally the positions of the bridges are adjusted so as to get maximum amplitude of vibration of the wire. Measure the distance between the bridges.
5. Repeat this determination 5-6 times and find out the mean value of l. As a next step the tension applied to the wire is altered and several sets of observations for l with different values of tension are taken.

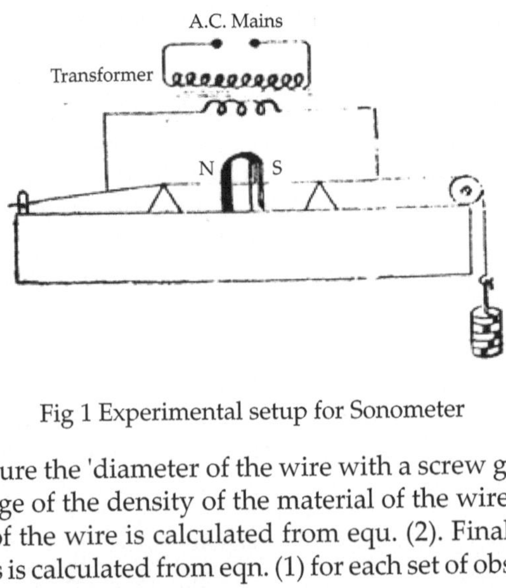

Fig 1 Experimental setup for Sonometer

6. Now measure the 'diameter of the wire with a screw gauge and assuming a knowledge of the density of the material of the wire. The mass per unit length m of the wire is calculated from equ. (2). Finally the frequency of A.C. mains is calculated from eqn. (1) for each set of observation separately and then the mean is taken.

Observations

[A] Measurement of l and T

TABLE 1

Set No.	Tension applied in kg	Length of wire between the bridges at resonance (cm)		Mean length l cm.	Frequency of A.C mains (Hz)	Mean frequency (Hz)
		As obtained by increasing the Length of the wire	As obtained by decreasing the length of the wire			

[B] Measurement of diameter of the wire

Least count of the screw gauge = ... cm.

Zero error (if any) = ... cm.

TABLE 2

Sl. No.	Reading along any diameter a cm.	Reading along perpendicular diameter b cm.	$\frac{a+b}{2}$ cm	Mean Diameter (cm)

Calculations

Density of the wire (given) = ... gm/cc

Mean corrected diameter of the wire = ... cm.

Radius of the wire = ... cm.

Mass per unit length of the wire $m = \pi r^2 \rho$ = ... gm/cm.

Set 1. $f = \frac{1}{2l}\sqrt{\frac{Mg}{m}}$ = ... = ... Hz.

Mean value of f = ... Hz.

Results

The frequency of A.C mains = ...Hz.

Sources of Error and Precautions

1. The horse-shoe magnet should be placed vertically at the centre of the wire with its face normal to the length of the wire.
2. The wire should be horizontal and pass freely in between the poles of the magnet.
3. The A.C. should be stepped down to about 6 or 8 volts before passing through the wire.
4. The sonometer wire and the clamp used to hold the magnet should be of non-magnetic material.
5. The wire should be uniform and free from kinks.
6. The tension applied to the wire should not stretch it beyond elastic limit (2.5kg).
7. The diameter of the wire should be measured at a number of points along its length and at each point two mutually perpendicular diameters should be measured.

Viva Voce

1. What do you mean by AC?
2. What is the standard value of frequency of a a.c. supply?
3. What is resonance?
4. What are damped and forced vibrations?
5. How does a transfer works? Explain.
6. What type of transformer is used in this experiment and why?

Experiment No. 27

Planck's Constant

Object

To determine the value of Planck's constant and work function of the material of the cathode of a photo-electric cell.

Apparatus

A vacuum type photo-electric cell enclosed in a light-tight box containing a high intensity mercury arc light source, pico-ammeter, d.c. voltage source 6V, light filters (red, yellow, green, blue, violet), key, connecting wires.

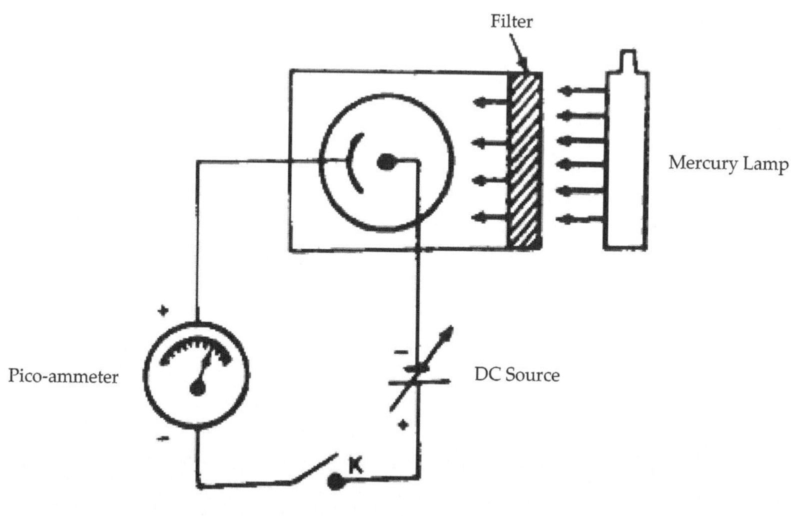

Fig. 1Experimental Setup for determination of Planck's constant

Principle

The schematic circuit diagram of Planck's experiment is shown in Fig. 1. When the light is incident through the filter onto the cathode, photoelectric emission begins and the current constituted by photoelectrons is measured in the pico-ammeter. The retarding potential, i.e., negative potential on the anode is slowly increased till the stopping (or cut-off) potential is reached reducing the current to zero. It is important to note that, due to some stray emissions, some time, the current may not actually reduce to zero but will acquire a small constant value (Fig. 2). Filters of different colours are used successively so that light of different frequencies are incident on the cathode. When the stopping potential is obtained for each frequency, a graph is plotted between stopping potential vs. frequency as shown in Fig. 3.

From Einstein's photoelectric equation, we have :

$$K_{max} = hv - W0$$

work function W0 = hvo, Where vo is the threshold frequency of the emitting surface. Also, Kmax = eVo, where Vo is the stopping potential.

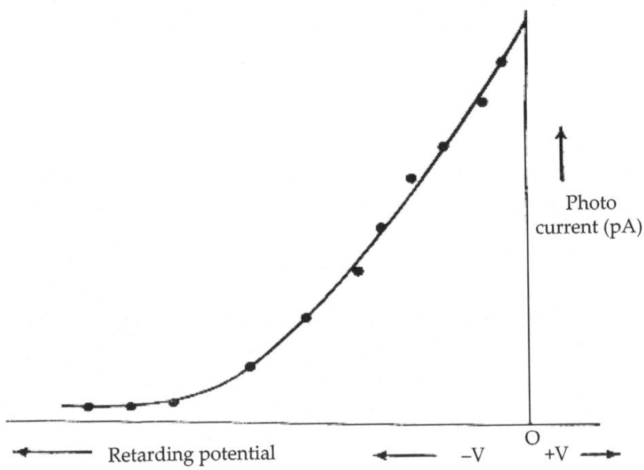

Fig. 2 Curve between retarding potential and photo current.

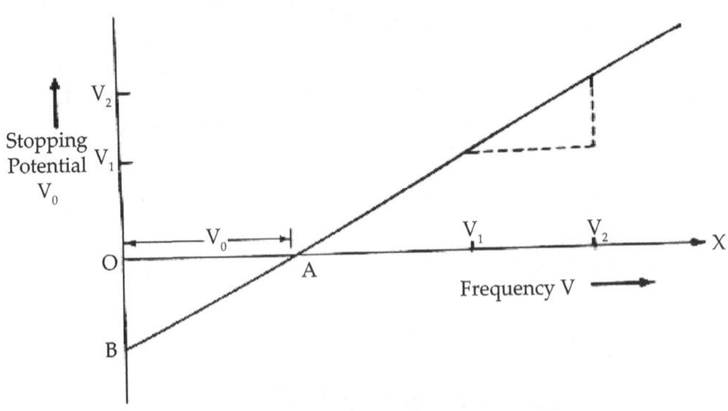

Fig. 3 Curve between frequency and stopping potential

$$eVo = hv - hvo$$

$$V_o = \frac{h}{e}v - \left(\frac{h}{e}\right)v_o \qquad (1)$$

This is analogous to the equation of a straight line y = mx+c,where slope m is equal to :

$$slops = \frac{h}{e}$$

$\therefore$ Planck's constant = e X slope (2)

Thus, by finding the slope of Vo vs. v curve we can calculate the Planck's constant using Eq. (2).Also, by noting down the intercept OB $\left(= \frac{hv_0}{e}\right)$ on the y-axis, we have.

Work function Wo = e . OB (3)

Procedure

1. First of all make the cicuit connections as shown in Fig. 1.
2. Switch on the instruments only after familiarizing yourself with all the instruments. Set the applied voltage zero and select scale in the pico-ammeter.
3. The mercury light source and the photo electric cell are adjusted so that light falls on the photocell window and the pico-ammeter reading is about 50 pico-ampere.
4. The violet filter is placed in front of the photoelectric cell in the socket and the current and voltage noted down. Now, the voltage in small steps and note down the current in the pico-ammeter is noted down. This process is continued till the applied potential becomes equal to the stopping potential and so that current acquires a minimum constant value.
5. The above process is repeated with blue , green , yellow and red filters.

Observations

TABLE 1

Sr. No.	Retarding Voltage (Volt) V	Photo-Current				
		Violet filter	Blue filter	Green filter	Yellow filter	Red filter
1.	0.0					
2.	0.05					
3.	0.10					

Plot the graph between photo current and retarding potential i.e., the potential when the photo current becomes nearly constant and minimum for all the filters (Fig.2). Then prepare the following table :

TABLE 2

Filter Wave length λ	Violet 4050 A°	Blue 4360 A°	Green 5460 A°	Yellow 5780 A°	Red 6910 A°
Frequency $v = \frac{c}{\lambda} Hz$					
Stopping potential V_o					

A graph is plotted between frequencies and stopping potentials (Fig.3) and its slope is calculted.

Shape = $\frac{V_2 - V_1}{v_2 - v_1}$ =.................volt/ Hz = ...volt sec.

Therefore, Plank's constant h = e.slope

= 1.6 x 10-19 C x ...... volt sec.

= ....Joule sec.

Work function Wo $= e.\frac{hv_o}{e}$

= e X OB Joule

= .... electron-volt.

Result

1. Planck's constant h = ... Joule sec.

   Standard value = 6.63 x 10^{-34} Joule sec.

   % Error = ...%

2. Work function W0 = ...eV

   Standard value =........... eV, % Error = ... %

Precautions

1. It should be ensured that when no light is falling on the cathode the reading in the pico-ammeter is zero for the applied retarding potential.
2. When the value of retarding potential for which the photo- current becomes zero or a minimum, at least three more readings must be taken by further increasing the negative potential in very small steps.

3. If not fixed, it should be ensured that distances between light source, filter and photo-cell remain constant during the experimentation.

Viva-Voce

1. What is pico-ammeter least count?
2. Explain Stray emission.
3. How does a photo-electric cell work? What is photo-electric effect? Explain.
4. What is retarding potential and stopping potential?
5. What is the standard value of Planck's constant? What is the application of Planck's constant value?

Experiment No. 28

E/M of an Electron

Object

To determine the value of ratio of charge E and Mass M (E/M) of an electron by Thomson's method.

Apparatus

In this apparatus, a Cathode ray tube fitted on a wooden stand is provided with two arms perpendicular to the axis of the tube and having scales with zero at the point where the central line of the arms cuts the axis of the tube. It is also provided with a power supply unit capable of giving the accelerating d.c. voltage for the CRT tube and the deflecting d. c. voltage for the electric field through a reversing key with suitable voltmeters, one pair of bar magnets, compass box, etc.

Description of Apparatus

Thomson's cathode ray tube, shown in Fig.1 consists of the following three essential parts:

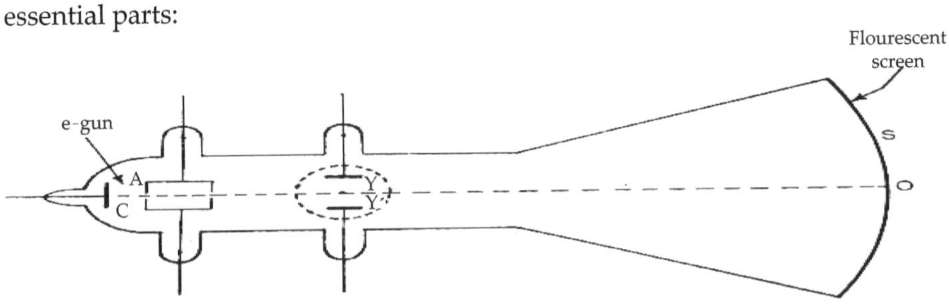

Fig. 1 Cathode Ray Tube (electron gun, deflection plates and fluorescent screen)

(i) The electron gun: The electron gun consists of the cathode C and the anode A. It produces, accelerates and focuses the electrons into a fine beam so as to produce a small bright spot on the screen S. Narrow slits are cut in opposite plates at A so that the cathode rays passing through are limited to a narrow beam.

(ii) The Screen: The screen S has a coating of some fluorescent materials like zinc sulphide, etc. so that a greenish blue spot of light is observed where the electron beam strikes upon the screen and interacts with the fluorescent material. The screen has a graduated X-Y (mm) scale on its surface to note the exact position of the fluorescent spot.

3. The electric and magnetic deflecting system: Two deflection plates YY' are fitted in the tube symmetrically on either side of the electron beam so that an electric field perpendicular to the plane of the paper can be applied between them. The cathode beam can also be magnetically deflected by placing a magnet on the wooden stand. When the bar magnets are placed with their lengths parallel to the arms with their opposite poles towards CRT, these bar magnets produce a uniform magnetic field at the same place where electric field is produced when potential difference is applied between plates (YY') as shown in Fig. 2. The deflection of the cathode beam due to both electric and magnetic fields takes place in a direction perpendicular to the plane of the paper.

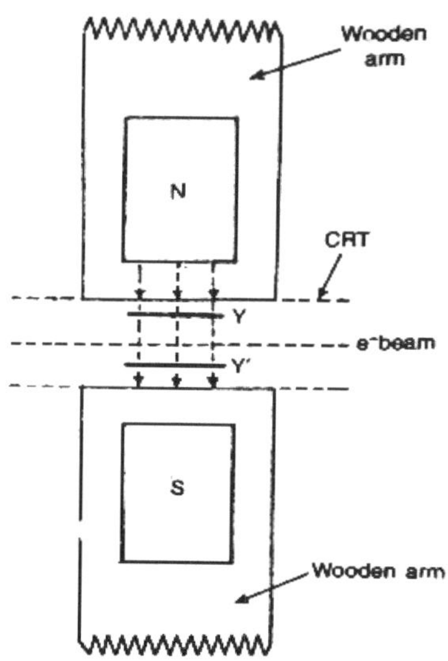

Fig. 2 Direction of electric and magnetic field in CRT

Theory

Let us suppose that E be applied-electric field between the plates YY' in a direction perpendicular to the plane of the paper then the force acting on an electron of charge e. Then

$$F_E = eE$$

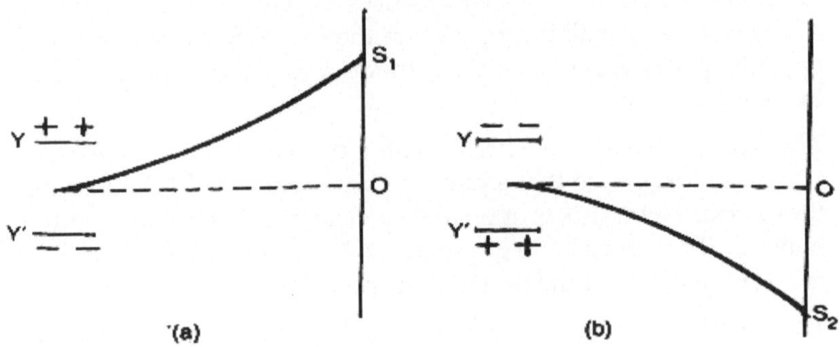

Fig. 3 Direction of the electric field

The direction of this electrostatic force is at right angles to the direction of motion of the electron. The electron, therefore, takes a circular path within the electric field and on leaving the field region moves tangentially hitting the screen at S1 or S2 depending upon the direction of the electric field, as shown in Fig. 3 (a) and (b).

When a magnetic field B is applied in the plane of paper perpendicular to the direction of the beam as shown in Fig. 3(a) then, according to Fleming's left hand rule, the beam is deflected upward out of the plane of paper in a direction perpendicular to both the direction of motion of the electron and that of the magnetic field. If the direction of the magnetic field is reversed, as shown in Fig. 3(b), then the electron beam is deflected downward into the paper. If v is the velocity of an electron, then the force acting on it due to the magnetic field :

$$F_B = Bev$$

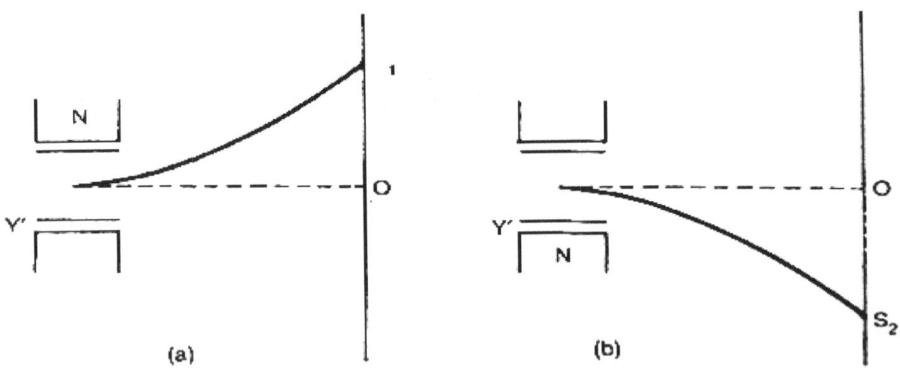

Fig. 4 Direction of magnetic field

The direction of this force is also at right angles to the direction of motion of the electron. Therefore, the electron takes a circular path inside the magnetic field and moves tangentially outside the field hitting the screen at S1 and S2.

If r is the radius of the circular path along which the electron moves in the magnetic field, then,

$$Bev = \frac{Mv^2}{r}$$

$$\frac{e}{m} = \frac{v}{rB} \qquad (1)$$

Determination of v: In order to determine v, one of the fields, (say the magnetic field) is first applied so that the spot of light on the screen moves to S1. Then the electric field is also applied simultaneously so as to produce deflection of the electron beam in the opposite direction. Now, the value of the electric field is so adjusted that the spot of light comes back to its initial position. In such a case, the forces acting on the electron due to the electric and magnetic fields are equal and opposite.

$$\therefore \text{Brv} = \text{eE}$$

$$v = \frac{E}{B}$$

Substituting for v in eq. (1), we get,

$$\frac{e}{m} = \frac{E}{rB^2} \qquad (2)$$

Determination of r: In order to determine the radius r of the circular path inside the magnetic field, the position of the bright spot is noted down in the absence of any field and again when the magnetic field is applied. Let the deflection of the spot be y = OS2.The circular path of the beam OR is as shown in Fig. 5. The beam leaves the magnetic field at R tangentially to the path and hits the screen at S2 so that S2R produces meets O'O at T. The perpendiculars O'C and RC to O'O and RS2 respectively meet at C. Therefore, C is the centre of the circular arc O'R and O'C = CR=radius of the circle = r.

Now,

$$\angle \text{O}'\text{CR} = \angle \text{OTS}_2$$

$$\therefore tan\theta = \frac{OS_2}{OT} = \frac{O'X}{O'C}$$

$$\therefore r = O'C = \frac{OT + O'X}{OS_2}$$

It is evident that, O'X is very nearly equal to the length of the region in which the magnetic field is applied. This is also equal to the length of the region in which the electric field is applied, i.e., the length of the plates Y and Y'. Let it bel, i.e., O'X = l ; this is normally given by the manufacturer in the catalogue. OT is the distance of the centre of the screen from the centre of the region of the magnetic field (or the electric field). Let it be L, i.e. OT = L,

$$\therefore r = \frac{Ll}{y}$$

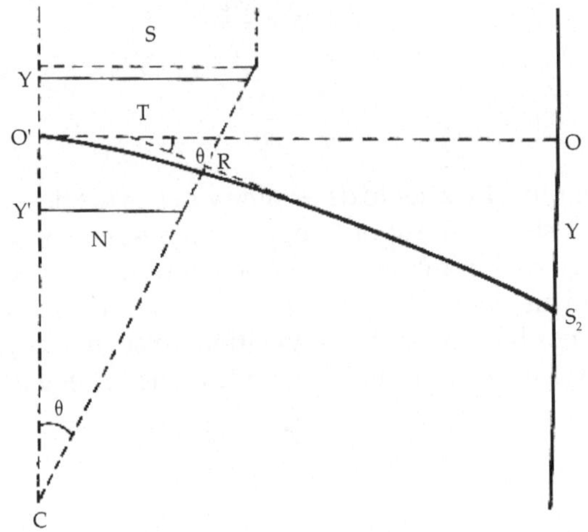

Fig. 5 Circular path of beam

Substituting this in equation (2), we get :

$$\frac{e}{m} = \frac{yE}{LlB^2} \tag{3}$$

The electric field E = V/d, where V is the applied potential difference between Y and Y'and dthe separation between the plates (given by the manufacturer). Substituting for E, we have :

$$\frac{e}{m} = \frac{yV}{LldB^2} \tag{4}$$

Thus, knowing V (Volts), B (Tesla), d, l, L and y (all in metres), E/M (Coulombs/ kg) can be calculated using Eq.4.

Procedure

1. The wooden stand is placed for mounting the magnet in a direction perpendicular to the magnetic meridian with the help of a compass box. Now, the cathode ray tube is placed perpendicular to the wooden stand and at its centre. The longitudinal axis of the C.R. tube is then in the direction of the horizontal component of the earth's magnetic field.
2. Carefully study the instructions manual pertaining to the operation of the cathode ray tube. Switch on the power supply. Adjust the intensity and focusing knobs of the cathode ray tube until a well defined spot appears on the screen. Adjust it to lie in the centre of the screen. Note this position of the spot on the marked vertical scale on the screen. This is the position of the spot when no potential has been applied to the deflecting plates.

3. Now, a certain steady potential difference V volt is applied across the deflecting plates and it is measured with an accurate high resistance voltmeter. The spot on the screen is found to be displaced vertically upwards. This position of the spot is noted down. The difference between the two positions of the spot gives the displacement y, for the applied potential difference of V volt.
4. Now two equal large bar magnets are placed in the F W. direction, one on each arm of the wooden stand along the marked scale with their proper poles pointing towards the cathode ray tube so as to reduce the electrostatic deflection. The magnets should lie symmetrically from the axis of the C.R. tube and should be displaced along the scales symmetrically so as to reduce the displacement of the spot to zero. The positions of the poles of the magnets nearest to the C.R. tube are noted down on the scale fixed to the wooden frame. Let these be r1 and r2 respectively. This adjustment implies that the field H produced by the magnets in their respective positions produces an equal and opposite displacement of the electron beam.
5. as a next step the magnets are removed. The applied voltage is reversed when the spot on the screen will be deflected vertically downwards. The position of the deflected spot is noted down and the displacement for the same applied voltage V volts is found out. The mean value of y1and y2 are calculated which gives the mean displacement y for the applied potential V volts.
6. The magnets again placed on the wooden stand in the direction required to reduce the deflection and as before, the poles of these magnets are arranged symmetrically and at such a distance apart that the field of sufficient strength to reduce the deflection to zero. Again the positions of the poles of the magnet nearest to the C.R. tube is noted down, on the scale of the wooden stand. Let these be r1 and r2 Switch off the power supply. The value of B to balance the electrostatic deflection may be determined in the following manner :

Remove the cathode ray tube from the wooden stand, which carries the magnets and place a compass box such that its centre lies on the common axis of the magnets. The pointer is adjusted at zero-zero mark. The bar magnets are placed with the same opposite poles towards the C.R. tube at the same distances r1 and r2, respectively as in the first part of the experiment when the deflection was reduced to zero. Observe the deflection of the pointer, reading both ends. Find the mean deflection θ1.Reverse the magnets pole and adjust their distances at r1′and r2', the distances of the poles when the deflection was reduced to zero. Again read both ends of the pointer and measure the mean deflection θ2. Take the mean of θ1 and θ2. Let this be θ.

If BH be the horizontal component of the earth's magnetic field, the field B due to magnets is given by :

$$B = B_H \tan\theta$$

This gives B to balance the effect of the electrostatic field E. The experiment is repeated by altering V and adjusting and measuring B to balance.

If the constants of the apparatus l, L and d are known and BH is known at the place of observation, using Eq. (4) we can calculate e/m.

Observations

I. Known Constants

Length of the deflecting plates, l = ………. cm = ……..m

Distance of the screen from the end of the plates, L= ………. cm = ……..m

Distance between the plates, d= ………. cm = ……..m

Value of BH = ….. oersted = ……Tesla

II. Determination of y and V.

Initial position of the spot. = ………. cm = ……..m

TABLE 1

<table>
<tr><th rowspan="2">S. No.</th><th rowspan="2">Applied Potential Difference V volt</th><th colspan="4">Deflected position of spod when V is</th><th rowspan="2">Mean deflection $\frac{y_1 + y_2}{2}$ cm</th><th colspan="2">When y is restored to zero</th></tr>
<tr><th>Direct cm</th><th>y1 cm</th><th>Reversed cm</th><th>y2 cm</th><th>Position of magnet on east arm. Distance of the nearest pole to C. T. Tube cm</th><th>Position of the other magnet on west arm. Distance of the nearest pole to C. cm</th></tr>
<tr><td>1
2
3
4
5</td><td></td><td></td><td></td><td></td><td></td><td></td><td>N. pole r_1
S. pole r_2</td><td>N. pole r_1'
S. pole r_2'</td></tr>
</table>

III. Determination of B:

TABLE 2

<table>
<tr><th rowspan="3">N =</th><th colspan="2">For y restored to zero</th><th colspan="2" rowspan="2">Deflection of the compass needle</th><th rowspan="3">Mean</th><th rowspan="3">Mean deflection $\theta = (\theta_1 + \theta_2)/2$</th><th rowspan="3">$B = B_H \tan\theta$</th></tr>
<tr><th>Position of the magnet on E arm</th><th>Position of the magnet on W arm</th></tr>
<tr><th colspan="2">Distance of the pole nearest to C.R. tube.</th><th>one end degree</th><th>other end degree</th></tr>
<tr><td>1</td><td>N. pole r_1
S. pole r_2</td><td>N. pole r_1'
S. pole r_2'</td><td></td><td></td><td>$\theta_1 =$
$\theta_2 =$</td><td></td><td></td></tr>
<tr><td>2</td><td></td><td></td><td></td><td></td><td></td><td></td><td></td></tr>
<tr><td>3</td><td></td><td></td><td></td><td></td><td></td><td></td><td></td></tr>
<tr><td>4</td><td></td><td></td><td></td><td></td><td></td><td></td><td></td></tr>
<tr><td>5</td><td></td><td></td><td></td><td></td><td></td><td></td><td></td></tr>
<tr><td>6</td><td></td><td></td><td></td><td></td><td></td><td></td><td></td></tr>
</table>

Calculations

(i) $\frac{e}{m} = \frac{y_1 V_1}{LldB_1^2}$ =.................. C/kg

(ii) $\frac{e}{m} = \frac{y_2 V_2}{LldB_2^2}$ =.................. C/kg

(iii) -----------

(iv) ----------

Mean /m = ... C/kg.

Result

The value of e/m of an electron	= .... C/kg
Standard value	= 1.76 x 1011 C/kg (=1.76 x 107 e.m.u./gm)
Percentage error	= ....

Sources of Error and Precautions

1. The cathode ray tube should be placed with its longitudinal axis in the direction of the horizontal component of the earth's magnetic field.
2. The spot on the screen must be adjusted to be sharp and intense. It should not be allowed to remain on the screen at a given position for a long time.
3. When the potential difference V is applied to the deflecting plates, its value must be so adjusted as to give the displacement Y in exact mms.

As the plates are not too far apart we can take the electric field E so produced as a uniform field and of an extent coinciding with the limit of the plates, i.e., we can neglect the end corrections. We assume the field BH as uniform in strength which is essentially not correct.

Viva Voce

1. What is E/M of an electron?
2. Explain the term electron gun. What is the role of control grid G?
3. Explain the working of YY′ plates.
4. How does a fluorescent screen converts the electron beam into corresponding optical energy?
5. What is thermionic emission of electron from cathode surface?
6. What are the applications of cathode ray tube?

Experiment No. 29

AC Measurements

Object

To determine the impedance of a given a.c. circuit.

Apparatus

A capacitor, an inductance, resistance, an a.c. ammeter, an a.c. voltmeter, a step-down transformer having a number of tapping in the secondary and flexible cord for connections.

Theory

Let us consider a capacitor of capacitance C, an inductance L and a resistance R be connected in series; and let an alternating e.m.f. E be applied across the combination as shown in Fig. 2. Then, if i virtual amperes is the current in the circuit, we have

$$\text{Impedance} = \sqrt{([L\omega - \frac{1}{C\omega}] + R^2)} = \frac{E_0}{i_0} = \frac{E}{I}$$

Or

$$\text{Impedance, } Z = \frac{\textit{emf as measured by an a.c. voltmeter}}{\textit{Current as measured by an a.c. ammaeter}}$$

Alternatively, the current i in the circuit may be measured by applying various known emfs., E in the circuit and a graph plotted taking E along the X-axis and the corresponding values of i along the Y-axis. This will come out to be a straight line as shown in Fig. 1 and its slope QR/PR with the i-axis will give impedance of the circuit.

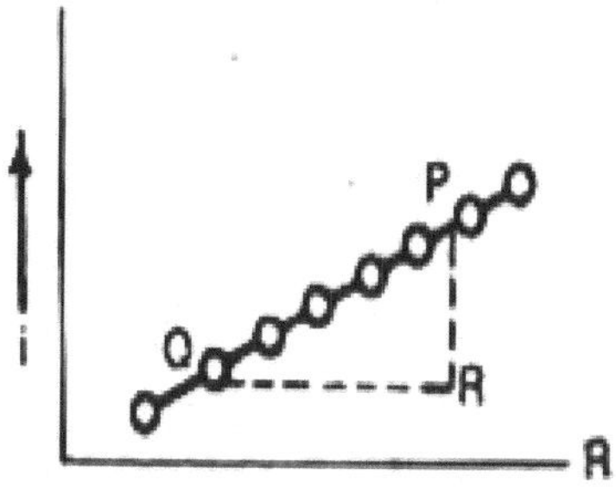

Fig. 1 Curve between current and emf

Theory

Circuit Containing Capacitance, Inductance and Resistance

When a harmonically varying emf E0 sin ωt is applied to a circuit containing a resistance R, an inductance L and a capacitance C in series as shown in figure 3.

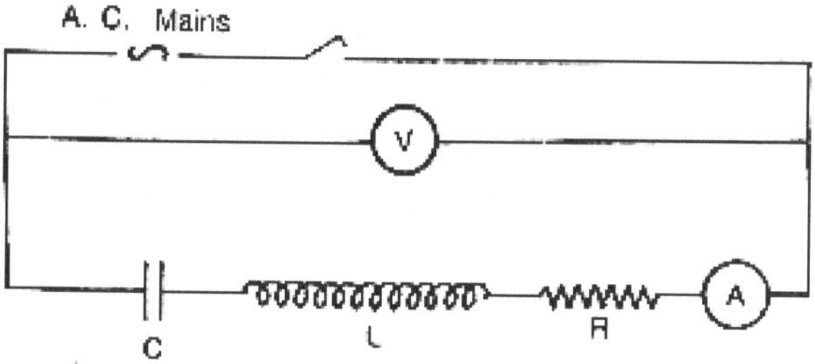

Fig. 3 Series combination of LCR

We have, $$Ri - L\frac{di}{dt} + V = E_0 \sin \omega\, t$$

Where I is the current and V the potential. Let q be the charge on the capacitor at the instant when its potential is V, then V= qC and we get from the above equation:

$$Ri - L\frac{di}{dt} + \frac{q}{C} = E_0 \sin \omega\, t \quad (1)$$

Let the solution be : i = A sin (ωt-a)

Where A and a are to be determined.

$$\therefore \frac{di}{dt} = A\omega \cos(\omega\, t - a)$$

$$\therefore dq = idt = A\sin(\omega\, t - a)\,dt$$

$$\therefore q = idt = \frac{A}{\omega}\cos(\omega\, t - a)$$

Substituting the values in eqn. 1, we get :

RA sin (ωt-a)-[L ω - 1/(ωC)]A cos (ωt-a) =E0 sin ωt

Comparing the coefficients of sinωt and cosωt on the two sides of the above equation, we get :

$$RA \cos a + [L\omega - 1(\omega C)]A \sin a = E0 \quad (2)$$

$$RA \sin a + [L\omega - 1(\omega C)]\, A \cos a = 0 \quad (3)$$

Squaring and adding eqns. (2) and (3), we have :

$$A^2\,[R^2 + \{L\omega - 1/(\omega C)\}^2] = E_0^2$$

$$\therefore A = \frac{E_0}{\sqrt{R^2 + [L\omega - 1/(wC)]^2}}$$

Also from eqn. (3)

$$\tan a = \frac{L\omega - 1/(\omega C)}{R}$$

Hence, $$i = \frac{E_0}{\sqrt{R^2 + [L\omega - 1/(\omega C)]^2}} \sin\left[\omega t - \tan^{-1}\left(\frac{L\omega / (\omega C)}{R}\right)\right] \quad (4)$$

The quantity Lω is the inductive reactance of the circuit and 1 /(Cω) the capacitive reactance. The total reactance of the circuit is the resultant of the inductive and capacitive reactances. The quantity $\sqrt{}$ Lω-1(ωC)^2+R^2 is called the impedance of the circuit and is denoted by Z. The total reactance of the circuit is denoted by X, the inductive reactance by XL and the capacitive reactance by Xc. It is clear from the expression for the impedance of the circuit that,

$$Z^2 = X^2 + R^2$$

From eqn (4) it is evident that the maximum value of the current is given by :

$$i_0 = \frac{E^0}{\sqrt{[L\omega - 1/(\omega C)]^2 + R^2}} \quad (5)$$

Thus, it is evident that the less the impedance of the circuit the greater is the value of the maximum current in it. The unit of impedance is ohm. After all, impedance is the effective resistance of an A.C. circuit. For a circuit in which Lω=1/ (ωC) is zero the current attains a value i_0= E_0/ R. Under such conditions the true ohmic resistance alone determines the current in the circuit.

The value of the angle a depends on the relative magnitudes of the inductive and the capacitive reactances in the circuit, ohmic resistance being constant. When Lω > 1/(ωC) the current lags behind the applied e.m.f., when Lω < 1/(ωC) the current leads the applied e.m.f. However, when Lω<1/(ωC) (the condition of resonance), the current is in phase with the applied e.m.f.

The vector diagram representing the p.ds. across the various components may be clearly and conveniently drawn. There will be three components to be laid off in the vector diagram.

(i) Ri the active component of the applied e.m.f along OA.

(ii) $L\omega i$ drawn 90 ahead of Ri, and

(iii) *i/(C)* drawn 900 behind Ri.

This is shown in figure 4.

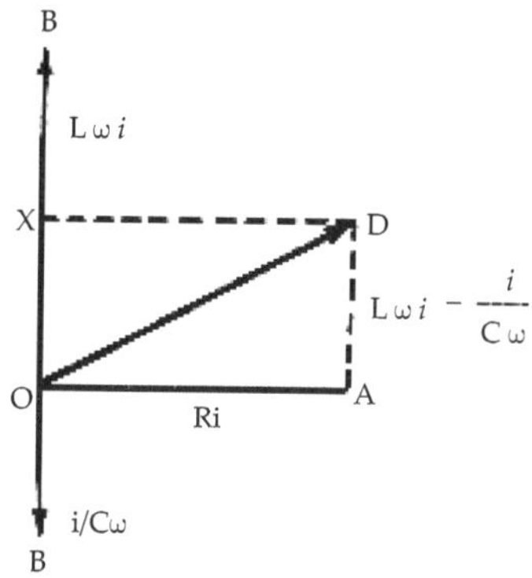

Fig. 4 Vector diagram

These three vectors can be combined by obtaining the resultant of any two and then combining this resultant with the third one. In the Fig. 4 $L\omega i$ and $i/(C\omega)$ have been first combined giving the resultant.

$$OX = \left(L\omega - \frac{1}{C\omega} \right) i$$

Which when combined with OA gives the resultant OD. Evidently,

$$OD = E = Zi = i\sqrt{R^2 + [L\omega - 1/(\omega C)]^2}$$

In the case illustrated, the resultant e.m.f is ahead of the current by the angle :

$$AOD = a = tan^{-1}\left(\frac{L\omega - /\omega C}{R} \right)$$ And the circuit behaves like a slightly inductive circuit. The current may be ahead or lag behind the impressed e.m.f. E according as $1/C\omega$ or $L\omega$ has the larger value.

Procedure

1. The primary of a step-down transformer is connected to the a.c. mains and connect to one of the tappings of the secondary a suitable capacitor, an inductance and a resistance, all in series, including an a.c. ammeter in the circuit.
2. As shown in Fig. 2, Connect a.c. voltmeter across the tapping of the secondary of the transformer to measure the virtual e.m.f. applied in the circuit. Having properly made the connections, the current is switched on and the reading of the voltmeter and the ammeter are noted down which give E and i respectively.

3. As a next step, after switching off the current, change the tapping of the secondary of the transformer and in this way take at least six sets of observations for i and E. The impedance of the circuit from each set of the observations are calculated separately and then the mean is taken.

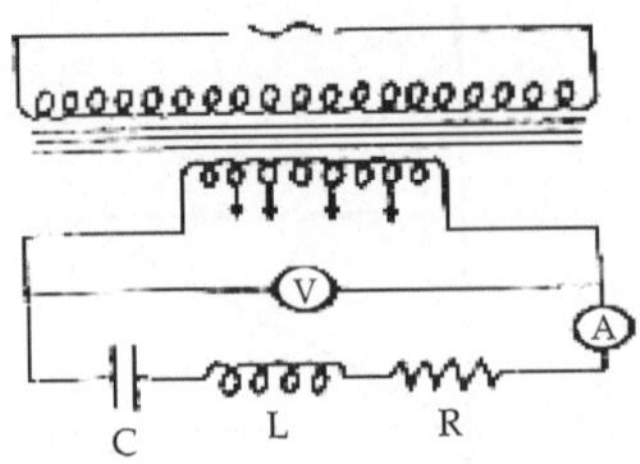

Fig. 2 Series combination of LCR

4. Next, a graph is plotted by taking applied e.m.f. E along the X-axis and the corresponding values of i along the Y-axis. This will come out to be a straight line. Measure its slope QR/PR with the i-axis which will give impedance of the circuit.

Also calculate impedance Z from the formula :

$Z=\sqrt{L\omega - \frac{1}{\omega C^2} + R^2}$, If the values of L, C and R are known.

Observations

L.C. of voltmeter = ... Volts.

L.C. of ammeter = ...Amp.

Sl. No.	Voltmeter Reading E (Volts)	Ammeter reading *I* (Amp.)	Impedance (ohm)
1. 2. 3.			

Calculations

$$Z=\frac{E}{i}=..........=...........ohms$$

[Make similar calculations for other sets]

Therefore, Mean Value of Z = ... ohm

From graph of figure 1

QR = ... PR = ...

Therefore, Impedance, $z = \frac{QR}{PR} = ...ohm$

Now, L=.....Henry, C=..... Fara

R=.....ohm and $\omega = 2\pi v$

$\therefore$ $Z = \sqrt{\left\{L\omega - \frac{1}{\omega C}\right\}^2 + R^2}$ = ... Ohm.

Result : The impedance of the circuit = ... ohm.

% error= ....

Sources of Error and Precautions

1. Both the ammeter and the voltmeter must be a.c. measuring instruments.
2. Connect the ammeter in series and the voltmeter in parallel in the circuit.
3. Zero error, if any, in the above measuring instruments should be taken into account.
4. The capacitor used should have a working Voltage greater than the a.c. mains voltage.
5. The inductance and the resistance used in the circuit should have a proper current rating so that they do not get over-heated and thereby get short-circuited and burnt.
6. Before changing the tapping of the transformer see that the current has been switched off.
7. The graph between i and E should be a straight line and should be smoothly drawn.

Viva Voce

1. Explain working of a step down transfer.
2. Explain the working of circuit elements R, L and C.
3. Explain the term electromotive force (emf).
4. How does a step up transformer & auto transformer works?
5. What are the applications of transformer & circuit components R, L & C?

Experiment No. 30

NPN Junction Transistor

Objective

To study the common emitter characteristics of an N P N junction transistor by drawing (i) input and (ii) output characteristic curves.

Apparatus

A N-P-N transistor, a voltmeter of range 0-3V, a voltmeter of range 0-10V, two microammeters of range 0-100 μA, a potentiometer of total resistance of the order of one mega ohm, a potentiometer of total resistance of the order of 25 kΩ, a 3V and a 9V battery, keys, connecting wires, etc.

Theory

Input Characteristics

Input characteristics of a NPN transistor are the curves showing variation of input (or base-emitter) voltage Vbe and input (or base) current Ib at different fixed values of collector-emitter Vce voltages.

A few typical input characteristic curves are shown in Fig. 1. From the curves, we can observe that as long as Vbe is less than the barrier voltage input current is low and when Vbe exceeds the barrier voltage,the base current increases first slowly and then rapidly similar to the case of a forward biased diode.

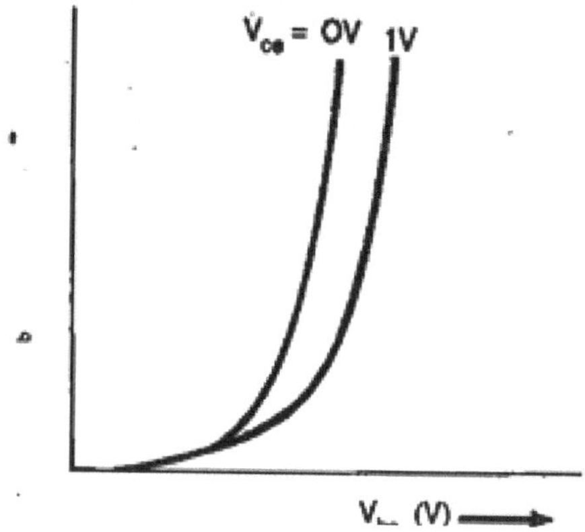

Fig. 1 Input characteristic curve

Output Characteristics

The output characteristics curve of a NPN transistor are the curves showing the variation of output (or collector-emitter) voltage V_{ce} and output (or collector) current I_c at different fixed values of base current I_b. A few typical characteristic curves are shown in Fig. 2.

From the curves shown in Fig. 1 and 2, we can observe that the collector current increases very rapidly with increasing Vce initially but soon it gets saturated and becomes almost independent of collector- emitter voltage. Further, for a given value of Vce, Ic is larger for larger value of Ib.

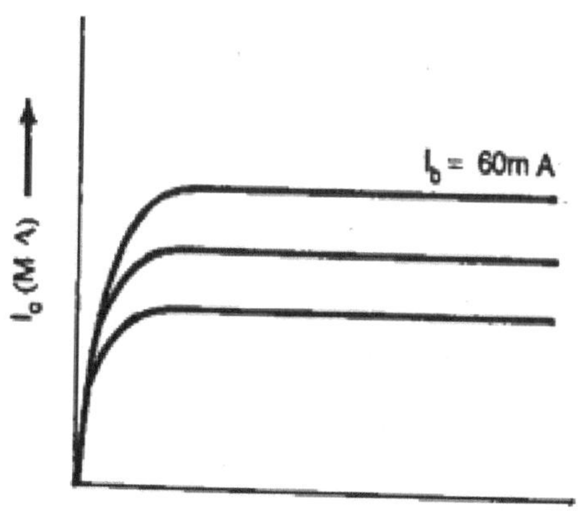

Fig. 2 Output characteristic curve

Working of N-P-N Transistor

For the conduction of a transistor, it is essential to apply suitable voltage or proper polarity to its terminals. For a transistor to work properly, the emitter base junction should be forward biased and the collector-base junction reverse biased.

The circuit diagram depicting biasing of a NPN transistor is shown in Fig. 3. In Fig. 4 its equivalent circuit using symbol of NPN transistor is depicted. The forward bias reduces the potential barrier and depletion width across the emitter-base junction. While the reverse biasing of collector-base junction increases the potential barrier and depletion width.

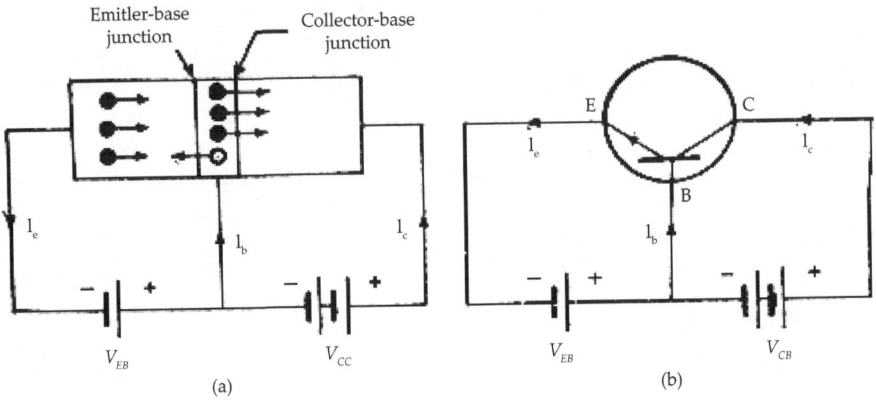

Fig. 3 N-P-N Transistor

Electrons in the emitter are repelled by the negative bias V_{EB} and the forward bias of the emitter-base junction helps the electrons in injecting into the p-type base region. Since the base is thin and lightly doped, only a small fraction (upto 5%) of the electrons recombine with the holes in this region and the remaining electrons diffuse into the collector. The positive bias of the collector pulls these electrons into the collector. The electrons flowing from the battery V_{EB} to the emitter constitute the emitter current , in the opposite direction. Some of the electrons entering the base from the emitter, recombine with the holes in the base and get neutralised. As the base region is very thin and lightly doped, the number of electrons recombining in this region is small- Moreover, due to a high positive potential at the collector, die electrons rush towards the collector. The electrons collected by the collector flow towards the positive terminal of battery V_{CB}. The electrons moving from the collector to the battery V constitute the collector current I.

The deficiency of the electrons inside the emitter is made up by the electrons flowing from negative terminal of the battery V_{EB}. Thus, the current is earned by the electrons both inside the transistor as well as in the external circuit. In the emitter-base external circuit the flow of electrons is from P to N. Hence, the conventional current is from N to P. In the collector-base circuit the flow of electrons in the external circuit is from N to P, resulting in conventional current I_c from P to N. The difference between I_e and I_{cis} the base current I_b which is in the same direction as that of I_c. Thus,

$$I_b = I_e - I_c$$

$$\text{or } I_e = I_b + I_c$$

$$\text{As } I_b << I_c,\ I_e \approx I_c$$

It is evident the total emitter current consists of a small fraction due to some holes diffusing from base to emitter. This total emitter current is very small compared to the current due to electrons crossing from emitter to base because of emitter being much highly doped as compared to that of the base. This ensures that the emitter current consists almost entirely of electrons. This is desirable since the current resulting from holes crossing the emitter junction from base to emitter does not contribute carriers which can reach the collector. Also, a small current called collector leakage current is constituted due to diffusion of some minority carriers (electrons) from the collector into the base.

Procedure

1. As shown in Figure 4 complete the circuit.

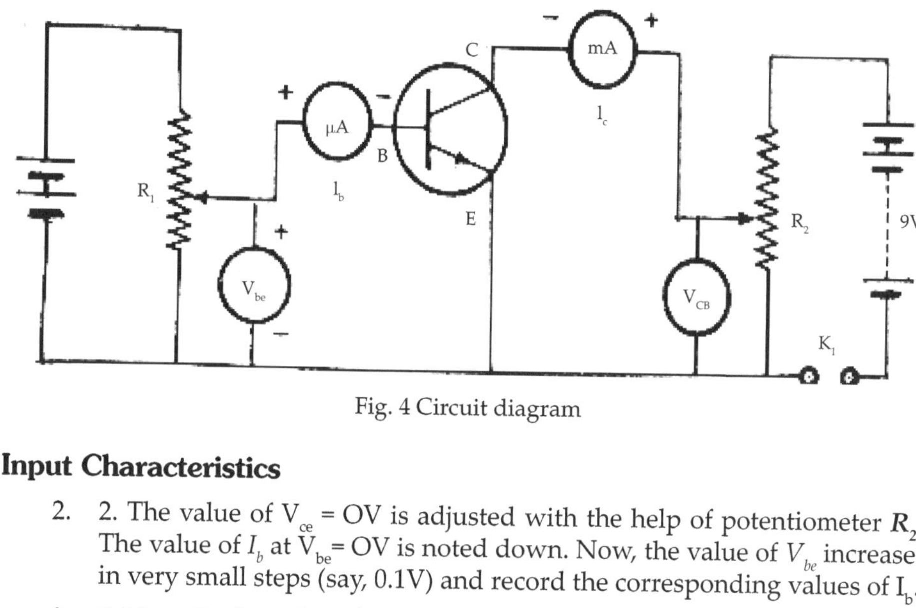

Fig. 4 Circuit diagram

Input Characteristics

2. 2. The value of V_{ce} = OV is adjusted with the help of potentiometer R_2 . The value of I_b at V_{be}= OV is noted down. Now, the value of V_{be} increased in very small steps (say, 0.1V) and record the corresponding values of I_b.
3. 3. Now, fix the value of V_{ce} = lV and keeping the value of V_{be} = OV record the value of I_b. As a next step increase the value of V_{eb} in small steps and record the corresponding values of I_b.
4. 4. The above process is repeated for the fixed values of V_{ce} =2V, 3V,......
5. 5. By taking V_{be} along the X-axis and I_b along Y-axis, the curve are plotted for various feed values of V_{ce} as shown m Fig. 1.

Output Characteristics

6. The value of I_b is adjusted at a fixed value say 20 μA, by using potentiometer R_1. Keeping the value of V_{ce} =0V the value of I_c is noted down. Now keeping I_b fixed increase the value of V_{ce} in small steps and the corresponding value of I_c is noted down.

7. As a next step, adjust the value of $I_b = 30\ \mu A$ and by keeping the value of $V_{cb} = 0V$ note down the value of I_c. Now keeping the value of I_b constant the value of V_{ce} is increased in small steps and the corresponding values of I_c are recorded.

8. The above process is repeated for fixed values of $I_b = 40\ \mu A,\ 50\ \mu A$ ....

9. By taking V_{ce} along the X-axis and I_c along the Y-axis, the curves for various fixed values of I_e, are plotted as shown in Fig. 2.

Observations

Input Characteristics

I_b Vs V_{be} at constant V_{ce}.

TABLE 1

Sl. No	Base-emitter voltage V_{be} (volt)	Input current I_b (μA)					
		$V_{ce} = 0V$	$V_{ce} = 1V$	$V_{ce} = 2V$	$V_{ce} = 3V$	$V_{ce} = 4V$	$V_{ce} = 5V$
1							
2							
3							
.							
.							
.							
10							

Output Characteristics

I_c Vs V_{ce} at constant Ib.

TABLE 2

Sl. No	Collector -emitter voltage Vce (V)	Output current Ic (mA)				
		Ib = 2 μA	Ib = 4 μA	Ib = 6 μA	Ib = 8 μA	Ib = 10 μA
1						
2						
3						
.						
.						
.						
.						
15						

Calculations

From the input and output characteristic curves obtain :

Input resistance r_i = ...

Output resistance r_o = ...

Sources of Errors and Precautions

1. Wrong polarity connection may damage the transistor and waste your time. Ensure proper connection of all transistor terminals.
2. Apply correct biasing potential in each transistor.
3. Excessive collector-base voltage may cause transistor breakdown, do not exceed the safe voltage limit.
4. Do not pass a large emitter current, it can burn the transistor.

Viva Voce

1. What are NPN and PNP transistors?
2. Explain CC, CE and CB transistors.
3. Explain the input and output characteristic curves.
4. What are the applications of transistor?
5. Explain the working of a transistor as oscillator.
6. Compare input and output characteristics of NPN transistor with PNP transistor.
7. Explain the biasing of a transistor.

Experiment No. 31

PNP Junction Transistor

Objective

To study the common base characteristics of a PNP junction transistor, by drawing (i) input characteristic curves and (ii) output characteristic curves.

Apparatus

A PNP transistor, a voltmeter of range 0-3V, a voltmeter of range 0-10V, two ammeters of range 0-25 mA, a potentiometer of total resistance of the order of one mega ohm, a potentiometer of total resistance of the order of 25 kΩ, a 3V and a 9V battery, keys, connecting wires, etc.

Working of Transistor

PNP Transistor

In this transistor also the emitter-base junction is forward biased while the collector-base junction is reverse biased, as shown in Fig. 1. In this case, majority carriers are holes and they are repelled towards the base due to positive potential applied on the emitter and the forward bias make the movement of holes across the junction easier. The base is thin and lightly doped so it has comparatively low concentration (or number density) of electrons. Therefore, only a fraction of holes (upto 5%) coming from emitter recombine with electrons and rest of the holes cross-over to the collector region.

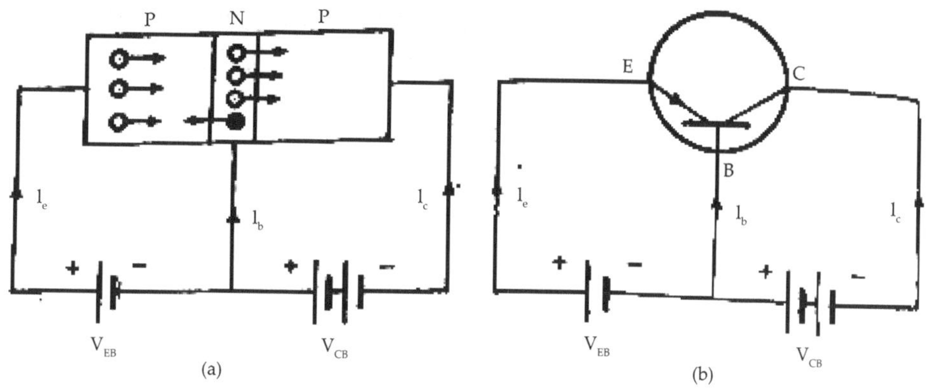

Fig.1 Experimental set up for P-N-P transistor

The reverse bias across the collector junction helps these holes in crossing over into the collector region. As one hole reaches the collector, an electron comes from the negative terminal of the battery V_{CB} and combines with it. At the same time, an electron from some covalent bond is released, which moves to the positive terminal of the battery V_{EE} and a hole is created in the emitter. Thus, inside a p-n-p transistor current is carried by hole while electrons carry current in the external circuit.

In the emitter base external circuit, the flow of hole is from P (emitter) to N (base), the conventional current I_e flows from N to P, i.e., into the emitter. At the collector, electrons enter into it from outside and, therefore, the conventional current flows out of the collector. The difference between I_e and I_c is the base current I_b. Thus,

$$I_b = I_e - I_c$$

$$\text{Or} \quad I_e = I_b + I_c \tag{1}$$

$$\text{As} \quad Ib << Ic, \; Ie \approx Ic$$

In this case also, the total emitter current consists of a small fraction due to some of the electrons diffusing from base to emitter. Also, a small current called collector leakage current due to diffusion of some holes from the collector into the base.

Theory

Input Characteristics

Input characteristics of a PNP transistor are curves showing the variation of input voltage or emitter-base voltage and input current or emitter current at different fixed values of collector-to emitter V voltages.

A few typical input characteristic curves are shown in Fig.2. Initially as input voltage is increased there is hardly any increase in the input current. When V_{eb} becomes greater than the barrier voltage, the current increases rapidly. The curve looks similar to characteristic curve of a forward biased *p-n junction diode*, because the emitter-base junction is forward biased. However, the base current is much smaller than the current that flows through the junction diode. Following points are evident from the curves :

(i) For a given collector-base voltage, the emitter current increases rapidly with increasing value of emitter-base voltage. This means that the input resistance is low.

(ii) At higher negative collector-base voltages, emitter current rises more rapidly with the emitter-base voltage. This is because an increase in the collector voltage results in a decrease in the base width, thereby decreasing the number of holes that recombine in the base region.

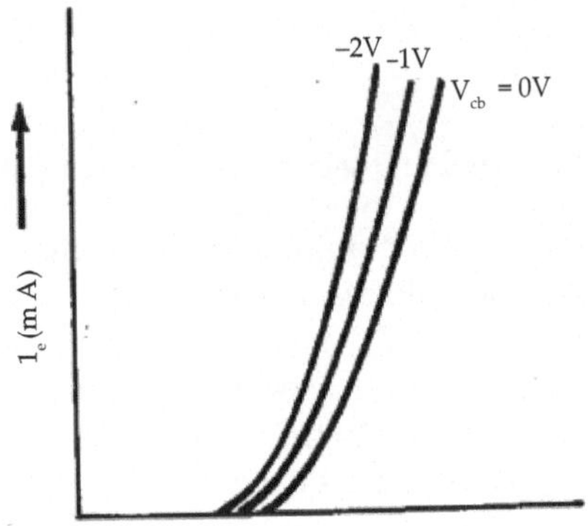

Fig. 2 Input Characteristics curve

The a.c. input resistance r_i of a transistor, in CB configuration, is defined as the ratio of small change in V_{eb} to the corresponding change in I_e at a fixed V_{eb}. That is,

$$r_i = \left(\frac{\Delta V_{eb}}{\Delta I_e} \right) \quad (2)$$

Output Characteristics

Output characteristics are curves representing the variation of output (collector) current I_c with output (or collector-base) voltage V_{cbat} different fixed values of I_e. A few typical output curves are shown in Fig. 4. The following points can be noted from the curves. : (i) For a given value of emitter current, the collector current is not zero when collector base voltage is zero, (ii) For a given emitter current, there is rapid increase in the collector current for an increase inflow collector-base voltage.

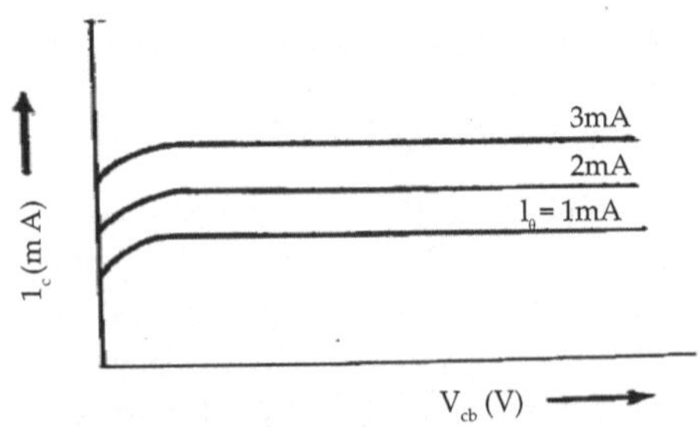

Fig. 3 Output Characteristics curve

In this region, the collector has low resistance.

(iii) For a given emitter current, the collector current becomes saturated beyond a certain value of V_b. This occurs when the collector voltage is sufficient high so that the collector collects almost all the majority carriers coming from the emitter-base region.

The a.c. output resistance r_0 is defined as the ratio of small change in V_{cb} to the corresponding change in I_c at a fixed value of I_e. That is,

$$r_0 = \left(\frac{\Delta V_{cb}}{\Delta I_e} \right)_{I_e} \quad (3)$$

Procedure

1. As shown in the circuit diagram in Fig. 4, complete the circuit diagram.

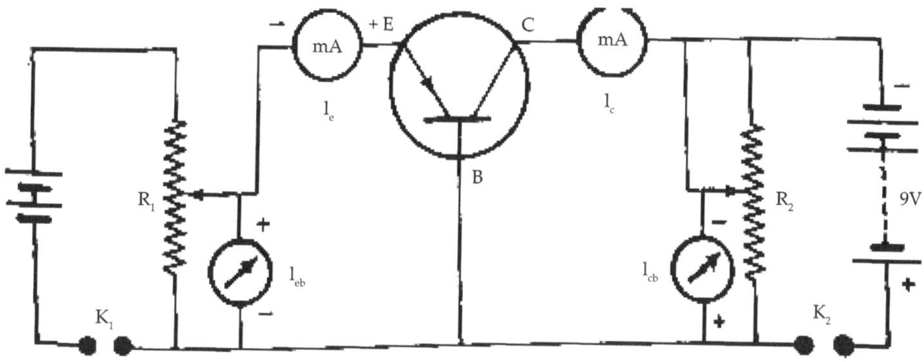

Fig. 4 Circuit diagram

Input Characteristics

2. The value of Vcb= 0 is adjusted with the help of potentiometerR2. Note down the value of 1 at Veb = 0. The value of V can be adjusted with the help of potentiometer R1. Now, the value of Veb increased in very small steps and the corresponding values of Ie are noted down.
3. Now, fix the value of Vcb = 1V and keeping the value of V =0 record the value of Ie. Now, increase the value of Veb in small steps and record the corresponding values of Ie.
4. The same process is repeated for the fixed values of Vcb= 2V, 3V,.........Taking Veb along the X-axis and Ie along Y-axis, the curve are plotted for various fixed values of Vcb, as shown in Fig. 3.

Output Characteristics

5. The value of Ie= 1 mA is adjusted with the help of potentiometer R1. Keeping the value of V_{cb} = 0 note down the value of I_c. Now, keeping I_e fixed, increase the value of V_{cb} in small steps (say, 0.1V) and the corresponding values of I_c is noted down.

6. Next, adjust the value of I_e = 2mA and keeping the value of V_{cb} = 0 note down the value of I_c. Now, keeping the value of I_e constant increase the value of V_{cb} in small steps and the corresponding values of I_c is noted down.
7. Now, repeat the above process for fixed values of I_e = 3 mA, 4 mA,.......
8. Taking V_{cb} along the X-axis and I_c along the Y-axis, the curves for various fixed values of I_e, are plotted as shown in Fig. 2.

Observations

(i) Input Characteristics

I_e Vs. V_{eb} at constant V_{cb}.

TABLE 1

Sr. No.	Emitter-base voltage Veb (volts)	Emitter current Ie{mA)				
		Vcb=0V	Vcb=1V	Vcb=2V	Vcb=3V	Vcb=3V

(ii) Output Characteristics

I_c Vs. V_cb at constant I_c.

TABLE 2

Sr. No.	Collector-base voltage V_{cb}(volt)	Collector current Ie{mA)				
		Ie=1mA	Ie=2mA	Ie=3mA	Ie=4mA	Ie=5mA

Calculations

From the input and output characteristic curves calculate :

Input resistance r_i = ...

output resistance r_o = ...

Calculate output resistance separately in the two regions, i.e., linear region and saturated region.

Sources of Errors and Precautions

1. Wrong polarity connection may damage the transistor and waste your time. Ensure proper connection of all transistor terminals.
2. Apply correct biasing potential in each transistor.
3. Excessive collector-base voltage may cause transistor breakdown, do not exceed the safe voltage limit.
4. Do not pass a large emitter current, it can burn the transistor.

Viva Voce

1. What is NPN and PNP transistors?
2. Explain CC, CE and CB transistors.
3. Explain the input and output characteristic curves.
4. What are the applications of transistor?
5. Name various types of transistor.
6. Expalin the biasing of a transistor.

References

A Text Book of Practical Physics Vol.1, ISBN: 81-225-0084-6, Kitab Mahal, Allahabad, Indu Prakash, Ram Krishna, A K Jha (2008).

A Text Book of Practical Physics Vol.2, ISBN: 81-225-0112-5, Kitab Mahal, Allahabad Indu Prakash, Ram Krishna, A K Jha, (2005).

Experiments in Engineering Physics (A Laboratory Mannual and Work Book with Viva Voce), ISBN: 81-219-1223-7, S Chand & Company Ltd. New Delhi, M N Avadhanulu, A A Dani, P M Pokley, (2009).

Manual for Experiments in Physics (With Equipment Developed for Indian Academy of Sciences, Bangalore) R. Srinivasan, K R Priolkar, T G Ramesh, (2015).

Laboratory Mannual, B. Tech. 1st Year Physcis, NIT Mizoram (2014).

Printed at Repro India Ltd.